Rare and Beautiful Minerals

Text by
Fritz Hofmann

Photographs by
Jürgen Karpinski

Rare and Beautiful

MINERALS

NEW YORK

Translated from German by Alisa Jaffa, Bromley
Revised by A C Bishop, London
English version by A R Woolley, London

Designed by Ludwig Winkler, Ilmenau
Manufactured by Druckerei Fortschritt Erfurt
Printed in the German Democratic Republic

Table of Contents

Minerals have fascinated man from earliest times. A great many of them have exceptionally attractive crystalline forms and beautiful colours, high refractive indices, and distinct hardness, clarity and lustre. It is these special and attractive properties of minerals that appeal to our feeling for beauty, arousing admiration and enthusiasm for their strange but perfect structure.

The rich variety of minerals presents the photographer with an almost inexhaustible diversity of subjects. Colour film is without doubt best for capturing the brilliant wealth of colour and the fascinating effects of minerals and, especially in the case of brightly coloured minerals, for showing clearly the gradations of colour that frequently occur even within a single mineral species. But even colourless minerals, the transparency of which reflects their purity or, in some cases, allows inclusions to be seen, are frequently reproduced in colour, as are ores with a metallic lustre or simply a grey appearance. The effect, however, may be to draw the eye to the liveliness of the background colour, thereby detracting from the delicacy of the mineral itself. With such subjects it is essential to capture their impressive formation in black and white. Skillful lighting can be used deliberately to highlight certain aspects of minerals from the overall impression. In this way black and white photography can achieve a likeness that is more faithful in terms of both form and colour, more valuable, exact and interesting than the superficial seductiveness of colour, quite apart from the fact that the expense of colour photography may not be justified by the result. A further consideration determining the choice between black and white and colour photographs was the desire to give some variety and balance to the book, and at the same time to single out the particular qualities of each of the minerals in a way that would do justice to their beauty and rarity.

Clearly, therefore, the most immediately striking feature of minerals is the beauty of their colour, but once the reader's attention has thus been captured, his interest will be further aroused by the fascination of their natural forms. Yet minerals do not only occur in the colours familiar to us, but can, as a result of association with, and inclusion of other, deeply coloured materials, range across the whole colour spectrum.

But the beauty of minerals is revealed not only by their colour, but also by their lustre and the natural association of different mineral species also contribute to their bewildering variety. In choosing minerals, it is important to examine them very closely to make sure they are in no way damaged, for the mechanical crushing techniques employed in mines and quarries frequently damage the specimens.

The attraction of most minerals lies in their exact geometric forms, which in nature result from the formation and growth of crystals. These processes are controlled by physico-chemical conditions, and result in an ordered arrangement of atoms that is characteristic of most minerals. The number of chemical elements which comprise minerals is comparatively small. It is thus all the more astonishing that so many beautiful and rare types of minerals can be formed from them. In this context, some rare minerals occur as a result of the unique presence of certain chemical elements in requisite stoichiometric proportions in the structure. The qualities of beauty and rarity have been preserved in a great many minerals, because they are so durable; although they formed millions of years ago, they have remained virtually unchanged.

It is these outstanding features of minerals that from earliest times have prompted man to adorn himself with coloured stones, and also to use them as weapons, implements, tools and raw materials. The origins of their names may be sought in antiquity. The historical development of the names of minerals, which are closely associated with their location, their characteristics and their use, but also with superstition and mysticism, will be examined in more detail later. The particular purpose of this book is to present rare and beautiful minerals from all over the world, the original specimens of most of which are to be found in the School of Mines in Freiberg.

Minerals may also be produced synthetically for use as jewellery and for technological purposes. This aspect is likewise dealt with in this book.

Explanation of Some Technical Terms

Crystals

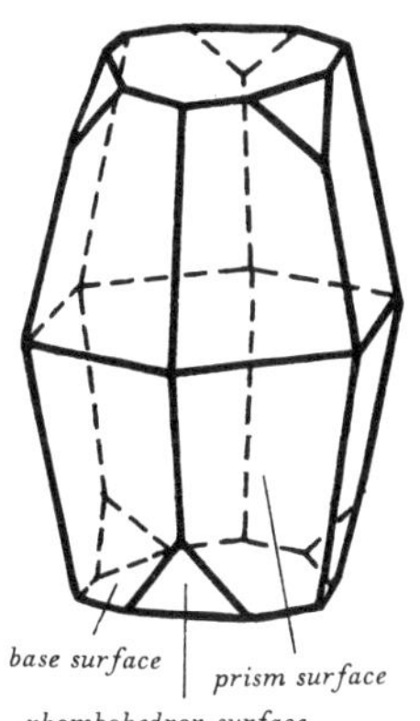

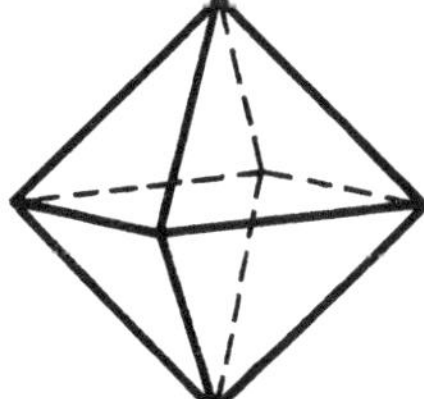

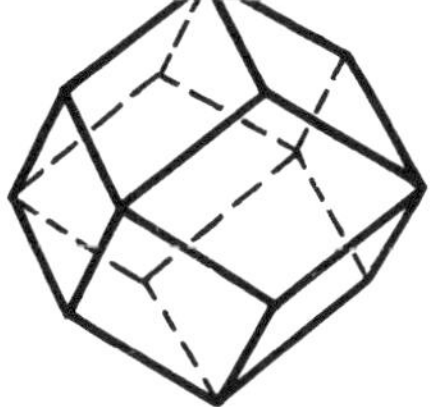

Simple crystal form—octahedron
Simple crystal form—rhombdodecahedron

In describing minerals, certain technical terms are used which will generally be unfamiliar. In an attempt at clarification, some of these are briefly explained here, not only to help the reader to understand minerals more fully, but also to help him to differentiate between minerals, rocks, ores and fossils and thereby define minerals without ambiguity.

A particularly interesting and impressive feature of minerals is the variety of crystal forms having a symmetrical arrangement. Let us begin by examining this aspect.

From time immemorial fine minerals and crystals have been regarded as something noble, precious or rare, and have captured the imagination of man. Minerals steadily acquired an increasing importance and interest, which in turn produced a desire to discover their nature, the causes of their different crystal forms and their splendid colouring.

In antiquity 'crystal' was known to the Greeks. They used the word *krystallos* meaning 'clear ice' to describe transparent rock crystal or quartz. A reference to it appears in PLINY: "At all events it is only to be found where winter snow brings extreme cold, and without doubt it is a form of ice." This notion of rock crystal being indissoluble ice did not persist for long, since rock crystals were found in countries which were free from frost. In historical descriptions, however, it is the characteristic and striking feature of 'crystal' that is stressed, namely the smooth faces by which it is bounded, and with the development of crystallography in the eighteenth century the term was extended to other minerals resembling rock crystal. This observation is still a part of the present-day definition of crystal, for it is a body bounded by smooth faces disposed in a regular arrangement.

In ideal conditions crystal faces are smooth planes which intersect along edges, which meet at corners. When the component faces of a crystal are of the same kind, they constitute a simple form, typical examples of which are the cube, octahedron and rhombdodecahedron. A crystal consisting of several simple forms is a combination, in which different simple forms may be equally developed, or one form may be better developed than the others and so becomes the dominant form of the combination.

Truncated corners and bevelled edges are characteristic features of a combination, and galena is a particularly good example, in which the dominant form is usually the cube. Its twelve edges are bevelled by the rhombdodecahedron, and the octahedron truncates the eight corners of the cube.

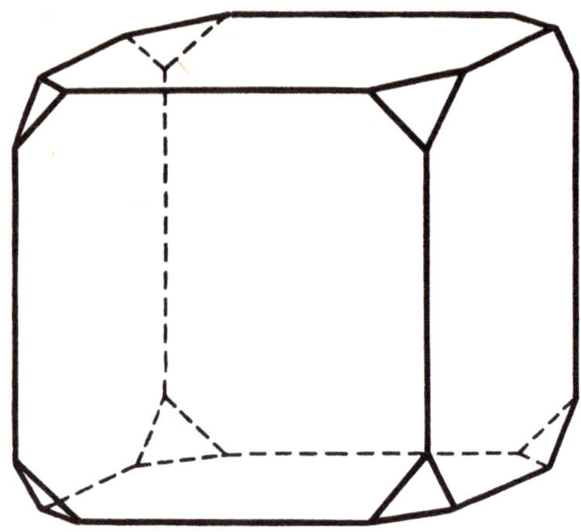

Combination crystal habit—cube with octahedron

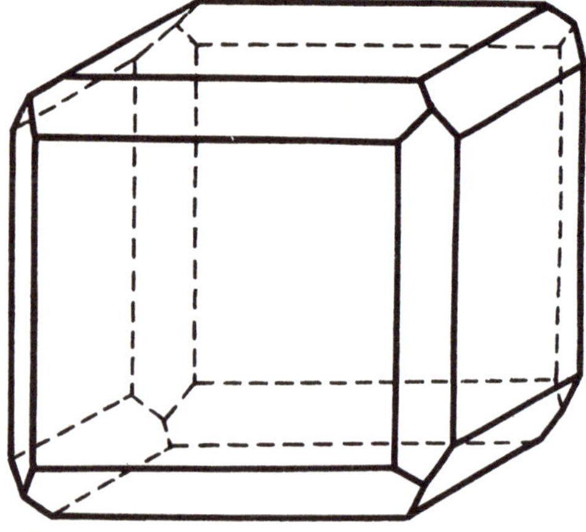

Combination crystal habit—cube with rhombdodecahedron

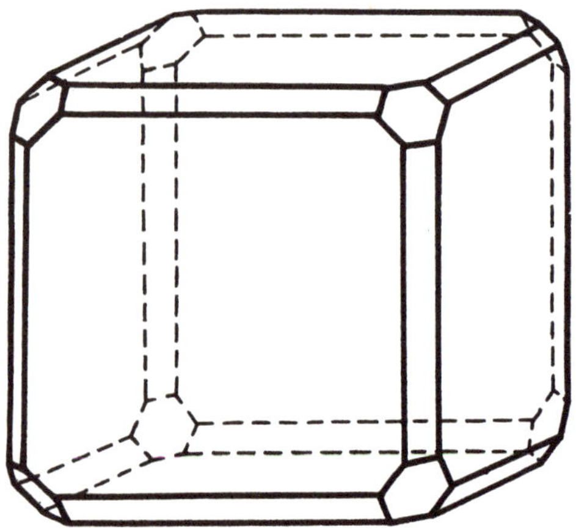

Combination crystal habit—cube with octahedron and rhombdodecahedron

Crystal faces belonging to the same form have a regular arrangement. Tihs feature is known as symmetry and is caused by the regular arrangement of the constituent atoms, ions, or molecules in the structure of crystals.

A plane of symmetry divides an ideal crystal into two equal halves, one of which is the mirror image of the other; for this reason it is also known as the mirror plane. A centre of symmetry requires that each crystal face has its corresponding parallel face with the same geometrical relationship to a central point, namely the centre of symmetry.

By rotating a crystal about an axis of symmetry, the chosen starting point may be regained after 60°, 90°, 120° or 180°. By turning the crystal through a full circle, it assumes a congruent position six times, four times, three times or twice. Axes of symmetry can, therefore, be six-fold, four-fold, three-fold or two-fold, and give to the crystal a general hexagonal, cubic, tetragonal, trigonal or rectangular shape.

The simple forms of the crystal, such as the cube, the octahedron and the rhombdodecahedron, are highly symmetrical forms, having a centre of symmetry as well as several planes of symmetry and two, three and four-fold axes of symmetry. There are very few examples in nature of crystals with a very low symmetry (one such is parahilgardite, a hydrated calcium chloroborate), because the low symmetry represents a corresponding state of order which must be correlated with a certain energy state. The underlying natural principle is for crystals to have the lowest possible energy state and this is attained in the familiar highly symmetrical crystal forms.

Crystals are divided into seven crystal systems with thirty-two classes on the basis of their symmetry. We do not intend to go into these here, but refer the reader to appropriate specialist works.

The growth of a crystal begins with the formation of its nucleus. For this to occur many molecules need to combine into an aggregate having a corresponding structure. If more of the same material is deposited on the outside, the nucleus continues to grow and a crystal is formed. In the course of its growth the different faces of the crystal do not develop at an even pace. The faces in an advantageous situation grow more quickly than others, and the shape of the crystal changes in consequence. Depending on the conditions in which they develop, crystals become either acicular (needle-like), columnar or tabular. If growth were even in all directions the crystal would be spherical. The general aspect of a crystal is known as its habit.

Some minerals have a wide diversity of shapes or habits. Calcite shows the greatest variety of habits: seventy-five different rhombohedra and more than two hundred scalenohedra have so far been observed. The opposite extreme, where a mineral presents very few habits or does not form good crystals, is rather rare.

The development of crystal forms can, as we have said, vary considerably in different mineral species, and depends on environmental factors. The most influential of these are temperature, pressure and the concentration of the solution from which the crystal develops.

In a natural crystal the growth of the faces is not necessarily uniform. Crystals frequently grow from a substrate (matrix) or from the walls of cavities or druses. As a result, the crystals will only develop to the full in one direction, remaining malformed or inhibited in others. Moreover, an ideal crystal form may be distorted by the incorporation of impurities in the structure and by changing concentration in the melt or solution from which it grows.

Minerals

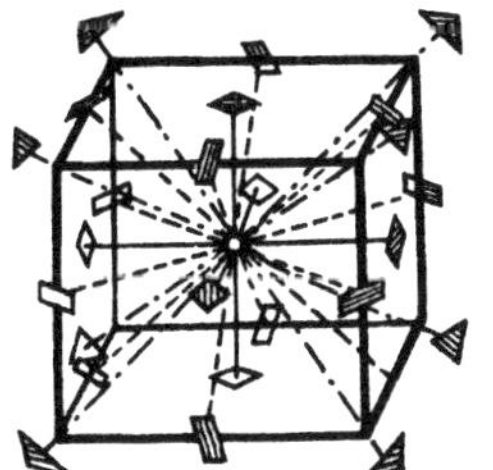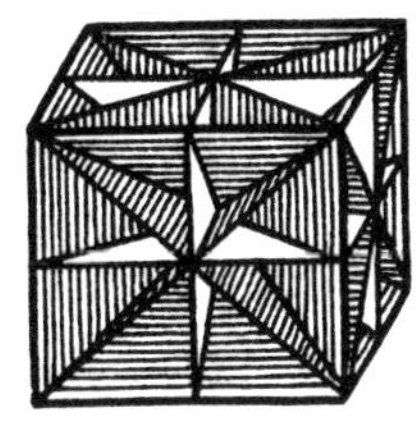

Axes of symmetry in cubic system
Planes of symmetry in cubic system

The word 'mineral' is derived from the French, *mine*, which came from the New Latin word, *mina*, meaning 'shaft', or 'mine'. Thus the origin of the word alludes directly to the place of extraction. In mines minerals were broken from ore veins, and the term, 'ore' was formerly and appropriately used to describe this material. It was only much later that this term was extended to include the actual ore mineral itself.

Minerals are defined as naturally occurring components of the Earth's crust having a definite chemical composition, that can be expressed as a chemical formula. When a mineral consists of a single element, the formula consists of the symbol for that element (e.g. gold or silver); in the case of compounds (e.g. oxides or silicates) the chemical formulae are made up of two or more symbols. Because, by definition, a mineral is a component of the crust of the Earth and occurs predominantly in a crystalline state, it excludes atmospheric materials, and even the water on the Earth's surface. Amorphous minerals, such as opal, are rare in comparison with the large number of crystalline minerals, and massive forms are more common than well-formed crystals.

Some minerals including both elements and compounds can crystallize in different modifications called polymorphs. Depending on the prevailing conditions, their constituent atoms will bond into different structures. Consequently the physical properties of the mineral polymorphs, for example the density or hardness, will differ even when the chemical composition remains the same. Carbon is a well-known example. The polymorph graphite is one of the softest minerals and has a perfect cleavage. Small cleavage flakes are flexible and have a greasy feel. These properties make it much in demand as a lubricant. Diamond, on the other hand, has an appearance that does not suggest carbon, while its exceptional hardness makes it an excellent material for drilling and cutting.

Calcium carbonate occurs in three modifications—as calcite, as aragonite and as vaterite. Pressure is an important factor in the formation of the different polymorphs of calcium carbonate: aragonite is stable at higher pressures than calcite, the commonest of these three polymorphs.

However, even when minerals have the same structure and almost identical chemical composition, they can vary considerably in colour and form. Such modifications are known as varieties. Quartz is a classic example. The best known coloured varieties are rose quartz (pink), citrine (yellow), blue quartz, amethyst (purple), smoky quartz (brown) and morion (black). The following are distinguished by their macroscopic appearance—sceptre quartz, skeletal quartz, and capped quartz.

Zinc blende likewise changes colour with the addition of iron. The pure mineral is honey yellow, the red variety resembles the precious stone ruby (ruby blende), apart from which there is the black variety, christophite. The names of minerals do not follow any fixed rules. Most have the suffix '-ite' or '-lite,' which derive from the Greek word '*lithos*,' meaning stone. Thus, for example, christophite is named after the St Christophe mine in Breitenbrunn in the Erzgebirge in the German Democratic Republic. Minerals have often been named in honour of their discoverers or of famous scientists, using either their first names or their surnames in recognition. Moreover, there are instances where the name derives from the colour of a variety of mineral, if this is particularly striking. Malachite and azurite owe their intense colours to copper, and the rich green and azure blue colour are one of their identifying characteristics. Many minerals may have characteristic colours, for instance cinnabar is red and sulphur yellow. In many minerals the colour results from the presence of strongly coloured substances. Some minerals notably labradorite and sunstone, show a spectacular iridescent play of colours. This is produced by diffraction and refraction of light by the surface or by cleavage planes, as well as by interference. This pseudochromatic colour is quite unrelated to the natural colour of the mineral.

The colouring materials or chromophores play only a subordinate role in the composition of the mineral. They are elements such as chromium, cobalt,

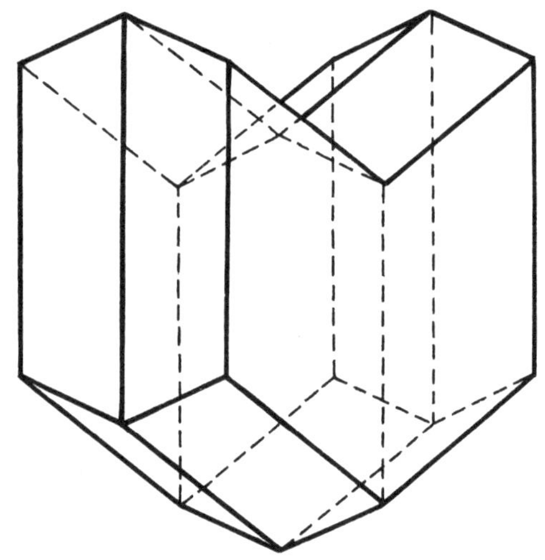

Gypsum—swallow-tail twin

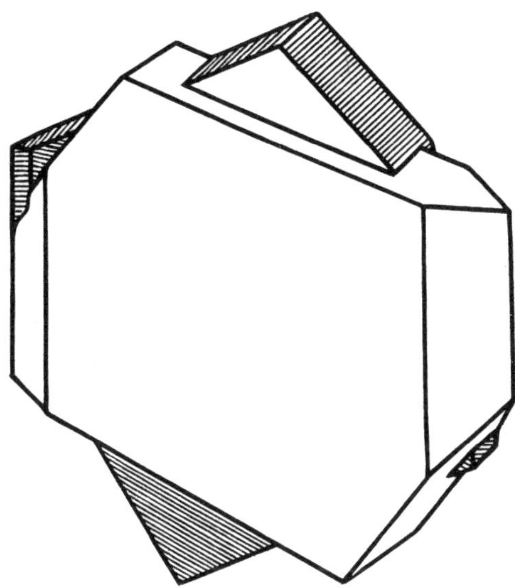

Twin of feldspar of potassium of Karlovy Vary

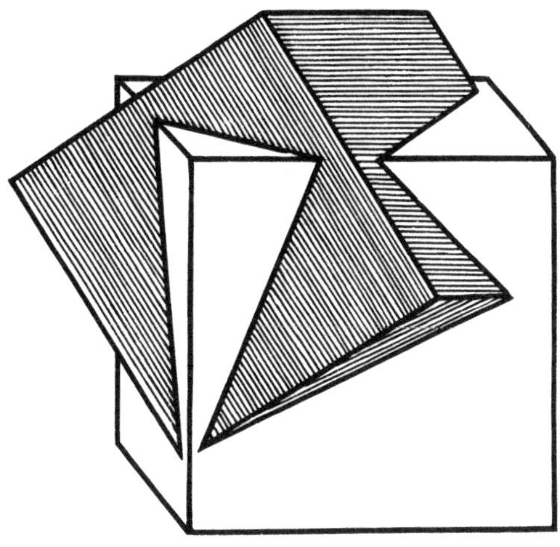

Twin of fluorspar

manganese, nickel and other heavy metals. The most common element in the structure of minerals is oxygen, followed by silicon, aluminium, iron, calcium, sodium, magnesium and titanium. The frequency of their occurrence in the composition of the known range of minerals is given below. Oxygen is a constituent of 1,221 minerals, silicon is present in 377, aluminium in 268, calcium in 134 and iron in 170. Though it should be pointed out, perhaps, that some 30–40 new minerals are described each year. Silicates and quartz are the most abundant minerals in the Earth's crust.

Minerals seldom form well developed isolated crystals. Crystals occur mostly as aggregates of several crystals, sometimes arranged in parallel groups. Crystal growth sometimes leads to 'twinning,' in which two individual crystals of the same mineral are in contact and related one to the other in a definite geometrical orientation. This twinning may be recognized by the presence of a re-entrant angle, or by the presence of a parallel band or bands, on the faces or cleavage planes of a crystal. The number of individuals comprising a 'twin' is not restricted to two—a twin may comprise three, four or more related individuals forming multiple or repeated twins. Twinning is characteristic of certain minerals, and may be used as a means of identification. The 'swallow-tail' twin of gypsum and the 'iron cross' twins of pyrite are typical examples.

Besides parallel growth between individual crystals of one mineral species, minerals commonly occur in association in mineral deposits or in rocks. Here, several minerals of differing structure and composition, which have developed under the same conditions, appear as a mineral association or paragenesis. For instance, in the southwest of England, minerals of tin, copper, lead, and zinc are found and have been mined since pre-Christian times. Similarly, lead and zinc minerals occur together in parts of the Lake District, Pennines and North Wales. These mineral associations are to be found not only in single deposits, but may also occur in extensive provinces, although all the mineralization within the province may not be contemporary.

The number of known minerals and their varieties is of the order of some 2,500. Between thirty and forty new minerals are described every year. On thorough investigation many of the new minerals have proved to be mixtures of two or more species, and have consequently not been confirmed by the Commission on New Minerals and Mineral Names of the International Mineralogical Association. The enormous industrial expansion of recent years has resulted in the demand for mineral raw materials far exceeding the natural supply. Synthetic production has to some extent helped to overcome this shortfall, but the artificial products are not minerals in the true sense of the word, although in the majority of cases their properties match those of natural minerals.

Ores

In the course of their formation, minerals can become locally enriched in the Earth's crust. Such concentrations are known as mineral deposits and, depending on the type of mineral, one or more metals may be extracted from the ores by smelting.

The Greek word *metallon* means 'mine' or 'pit.' By association, the meaning was later extended to include ore, as well as the metal sought and extracted from the mine.

In the past, the word metal was used much more commonly than ore, and even in the eighteenth century, A G WERNER still classified what we now know as ores, as metals. In his time, ores were still regarded as mineralized metals which had lost their lustre and their easy fusibility through the admixture of sulphur, arsenic or another metal.

Up to the Middle Ages there were seven known metals—gold, silver, copper, lead, tin, iron, and mercury. Since antiquity these had been linked with the planets, the Sun and Moon, Venus, Saturn, Jupiter, and Mars. Of these metals only mercury, being a fluid metal which could not be hammered, was like sulphur regarded by the alchemists as a raw material, and the planet Mercury was ultimately named after it. Even the early mineralogists, such as A G WERNER, used these terms when referring to metals, using the symbols of the planets in place of the names of the metals. The system of linking metals with the planets soon proved inadequate, however, when there were no longer enough planets available to fit the subsequently discovered metals, such as zinc, bismuth and cobalt.

Ores are divided into different types depending on their composition, in the same way as minerals, which occur either as the native element or as compounds. Precious metals occur naturally partly as the native element. The essential characteristic property of gold and silver is that they rarely form compounds, especially with oxygen.

Apart from gold and silver, the native elements play a subordinate role as ores. With the large scale extraction of heavy metal compounds nowadays native precious metals are becoming less important. Sulphides and oxides are considerably more important as ore minerals.

Sulphide ores can be conveniently grouped according to their general appearance. One such group consists of the minerals pyrite, arsenopyrite, chalcopyrite

Simple crystal form—pentagondodecahedron
Simple crystal form—diakisdodecahedron, diploid

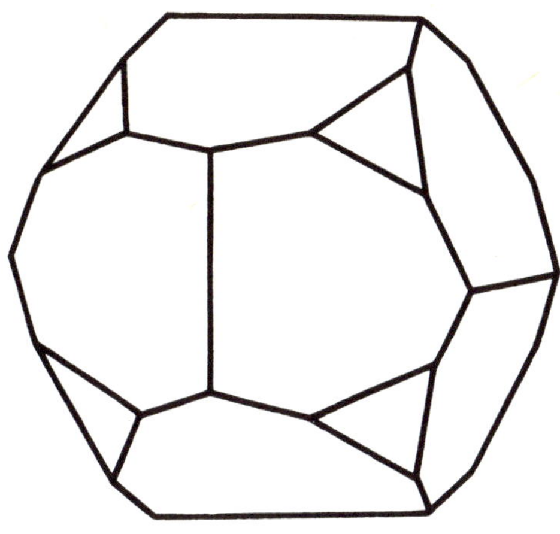

Combination crystal habit—pentagondode-cahedron with octahedron

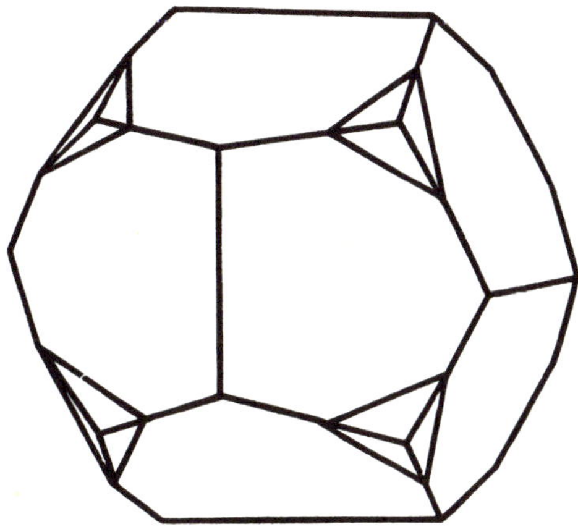

Combination crystal habit—pentagondode-cahedron with diploid

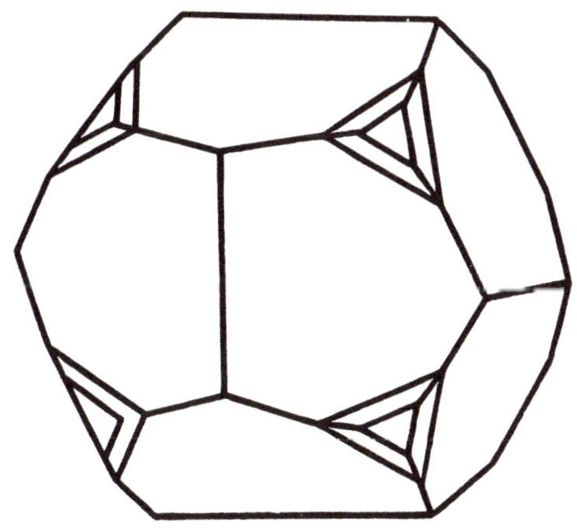

Combination crystal habit—pentagondode-cahedron with diploid and octahedron

and pyrrhotine, and they are characterized by their brilliant light colour and high hardness.

The composition of pyrite—iron and sulphur with small amounts of copper, silver and gold—has long been known, as well as its considerable hardness, which causes it to give off sparks when struck with steel. Despite its iron content, it is to this day not used as iron ore but is exploited for use in the manufacture of sulphuric acid.

Galena or lead glance, the lead ore commonly encountered in silver ore veins, has been known since the sixteenth century and is one member of another group of ore minerals. Its silver content varies between 0.2% and 2%, making this lustrous, cubic lead ore also an important silver ore. By contrast with the first group of sulphides, referred to above, galena was found by the old miners to be a good, useful ore which was easy to smelt.

The essential properties of this second group of sulphides is their brilliant metallic lustre, which is largely caused by the smoothness of the surfaces and the presence of a good cleavage. Other 'glances' are stibnite (antimony glance), molybdenite, argentite-acanthite (silver glance) and chalcosine (copper glance).

A third group of ores is characterized by a semi-metallic appearance, high density and low hardness, which led the early miners to think they could be easily smelted. The main member of this group is zinc blende (sphalerite) which, it was discovered in the Middle Ages, did not yield metal when smelted. Consequently, the German miners considered themselves to have been 'blinded' (from the German 'blenden') or hoodwinked and so referred to this ore as 'blenden.' It was only with the discovery of zinc in 1734 by the Swedish chemist, G BRANDT, that the name lost its original meaning. Antimony blende (kermesite), and arsenic blende (orpiment) have similar physical properties to zinc blende.

It was not only zinc blende that deceived the miners, but also pitchblende, a black, lustrous ore, from which the smelters found it impossible to extract any useful metal. In 1789, M H KLAPROTH discovered a new type of metal in the ore. By association with the planetary names, he called it uranium, after the planet Uranus which had been discovered in 1781. Pitchblende is the only 'blende' that is not a sulphide but an oxide.

Nowadays sulphides are no longer grouped according to their properties as was the custom with mineral ores, but this economically important group of minerals is classified according to the metal present and the ratio of metal to sulphur in the structure.

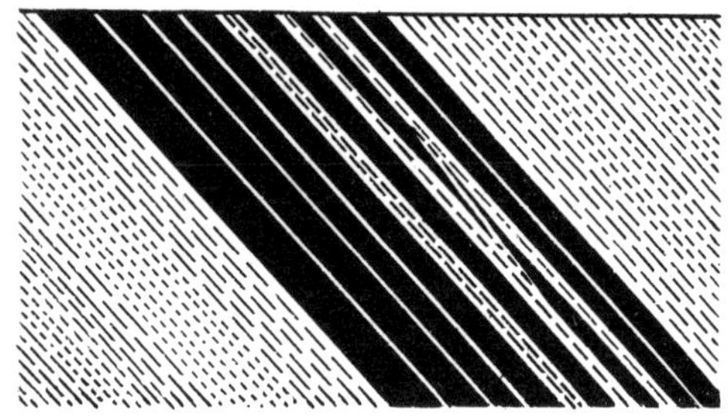

Structure of seams (in cut)

Rocks

In order to extract the metals from sulphide ores, they are subjected to intense heat, a process known as roasting, and thereby converted into the metallic oxide. Then an appropriate reducing agent, usually coke, is used to extract the metal from the oxide. Sulphur is driven off in the form of sulphur dioxide and the purified gas is the raw material in the production of sulphuric acid.

Apart from sulphides, the oxides of several heavy metals are of major importance as ores. This type of ore is composed of the metal, such as iron or manganese, in combination with oxygen. The oxides can be smelted without any major preparatory technical processing, such as are required for the sulphides.

The iron oxides are the best known of the oxide ore minerals. Magnetite is of particular importance, its iron content (70%) making it highly sought after. The iron content of the red iron ore, hematite is very nearly as high, followed by limonite with 55%–60% iron content.

Silicate ores are of subordinate importance and the extraction of metals from them is costly because of the complexity of the processes involved. In this group, the nickel ores, garnierite and schuchardtite are mined and smelted.

Man has used rocks for many purposes and their use reaches back into his early history. In the stone age rocks were used as tools, later as writing implements, and up to the present day rock retains its importance in architecture and in sculpture.

Rock occurs everywhere and in many forms. Mountain ranges consist of different types of rocks, and the recognition and nomenclature of rocks falls in the fields of petrography and petrology. The scientific study of rocks is petrology, and petrography is concerned with minerals, describing the mineralogical constitution of rocks.

Some rocks and many minerals have qualities which make them attractive so that they are regarded as gems to be worked into jewellery. In this context the term 'stone' has acquired a special meaning, quite distinct from its everyday use. In the jewellery trade the term 'stone' is used for cut crystals, whether they are precious or semi-precious varieties.

Let us now turn to the limiting definition of rocks used in terms of petrography and petrology, and the principles applied in classifying them. Rocks are natural aggregates of minerals, and whether they occur in a consolidated or an unconsolidated state, the mineral assemblages can be the same over large areas. Rocks are thus assemblages of minerals, but for the most part, only a small number of minerals make up the majority of rock types. Some rocks are composed pre-

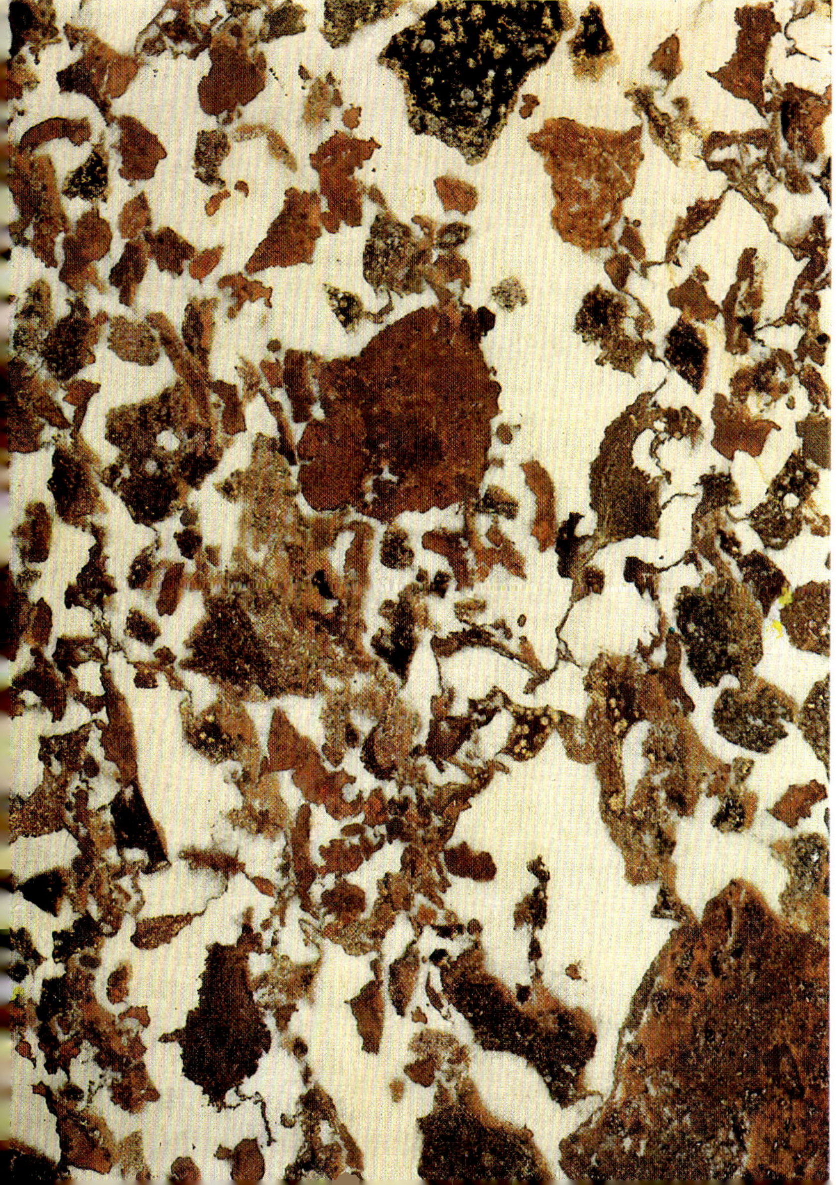

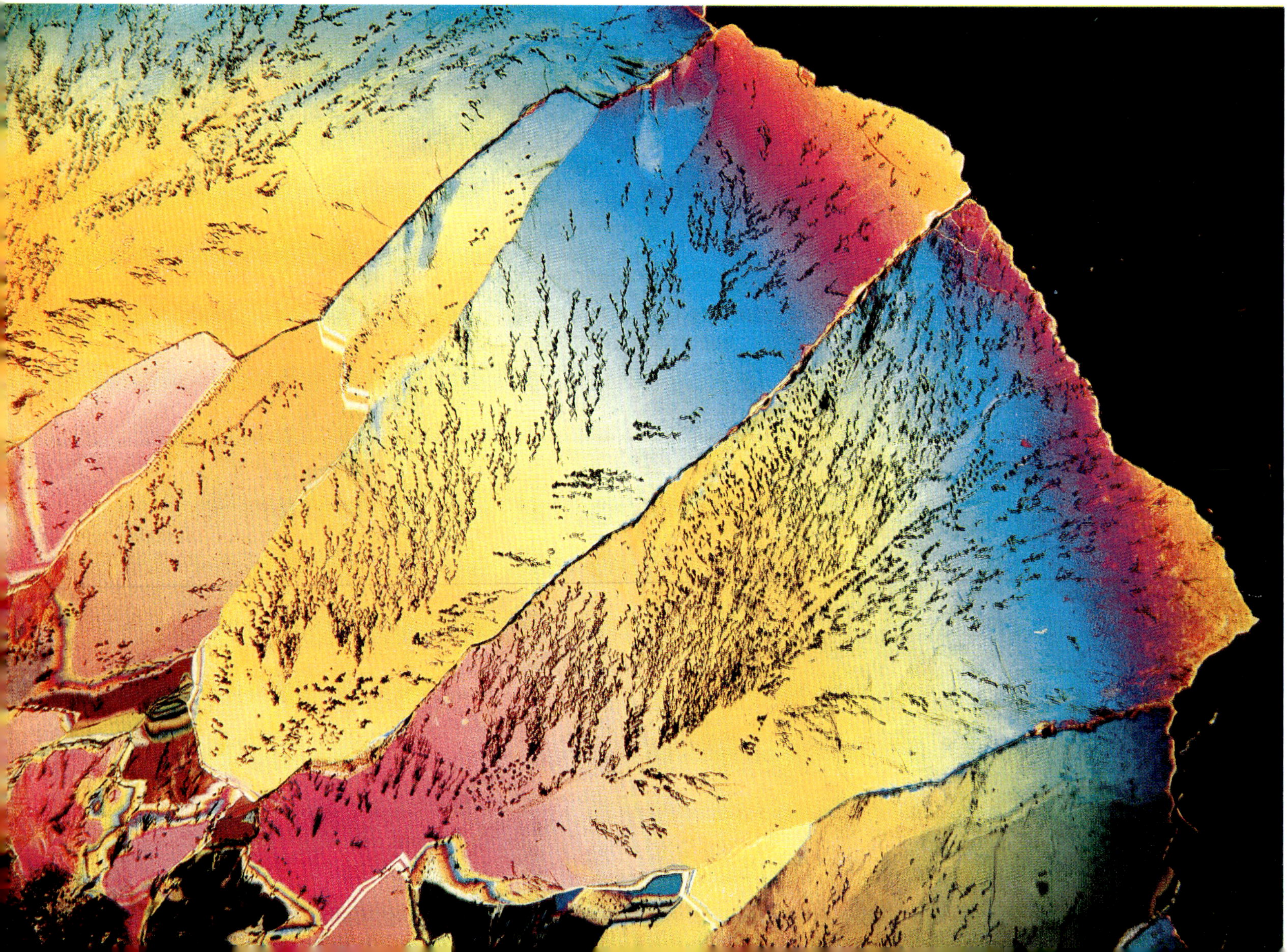

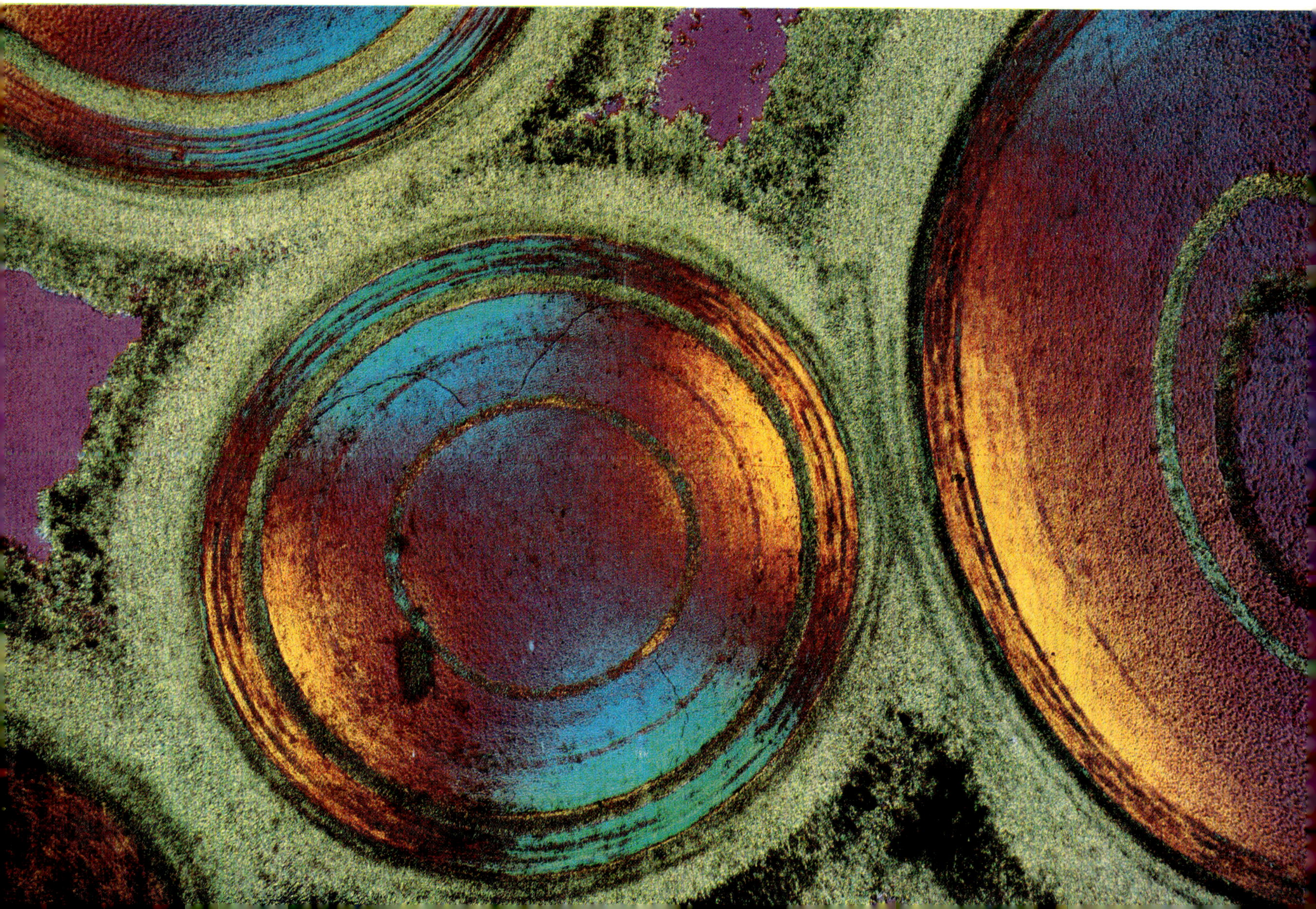

Graptolite, Monograptus (spirograptus)
spiralis spiralis (Geinitz), from the Valen-
tian: Silurian.
Grobsdorf near Ronneburg (G.D.R.)
Length: 7cm

Trilobite, Conocoryphe sulzeri (from
Schlotheim), middle Cambrian
Beroun near Prague (Czechoslovakia)
Length: 8cm

Outcrop in a sandpit
Brickwedde near Ankum (Federal Republic
of Germany)
Detail

22

Fish, Clupea sp., Eocene
Green River, Farson, Wyoming (U.S.A.)
Length: 9cm

Ammonite, Placenticeras placenta Dekay,
upper Chalk (Senonian)
Badlands, North Dakota (U.S.A.)
Diameter: 19cm

Lode

*Positions and band textures of the ore
(schematically represented)*

*Cockade texture of the ore
(schematically represented)*

dominantly of only one mineral; examples are limestone, dolomite and quartzite.

A uniform appearance may be imparted to a rock not only by the uniformity of the mineral assemblage, but also by the relative size of the individual mineral grains, which depends on the conditions obtained during the formation of the rock. Variation in grain size is one of the criteria used for differentiating between some types of rocks.

One of the principle groups of rocks are formed by the crystallization of a hot, silicate melt called magma and consequently they are called magmatic or igneous rocks.

Some igneous rocks crystallize within the Earth's crust and they cool only slowly. As a result, the mineral grains have time to grow relatively large, so such rocks, called intrusive rocks, tend to be coarse grained. Perhaps the best known example of the intrusive igneous rocks is granite. Over the centuries the meaning of granite has changed. In the sixteenth century a coarse-grained marble was known in France and Italy as granite, on account of its homogeneous texture. The name may be derived from the Latin, *granum* (grain). The essential constituents of granite are feldspar, quartz and mica.

Magma may form very large intrusions within the crust, or it may form narrow veins, perhaps in joints or fissures in the surrounding rocks. Around intrusions of granite, such veins are common and are referred to as pegmatites or aplites. Granite aplites are mostly fine-grained and pale in colour, whereas pegmatites are always coarse-grained. Perfectly formed crystals of varying sizes are commonly found in pegmatites. Large crystals of topaz, beryl, apatite, tourmaline, feldspar (orthoclase, albite), quartz and many other minerals are not uncommon in such rocks and are used commercially.

Magma can also rise up through the overlying rocks to the surface of the Earth and flow out over it giving rise to the igneous rocks known as volcanic or extrusive. These lava flows cool rapidly and generally contain only very small crystals. However, sometimes larger crystals do occur and these are often well formed. They are referred to as phenocrysts, and a rock containing large crystals set in a finer grained matrix is known as porphyritic. The name porphyry derives from the Greek, *porphyra*, meaning purple snake. The deep red or purple colour is found in only a few types of porphyry, but came to be applied to the texture as a whole, so that by association porphyry, or, more commonly, the adjective porphyritic is nowadays used for any igneous rock containing phenocrysts. The most abundant volcanic rock is basalt. In antiquity its colour and hardness were compared with those of iron.

Graphic granite

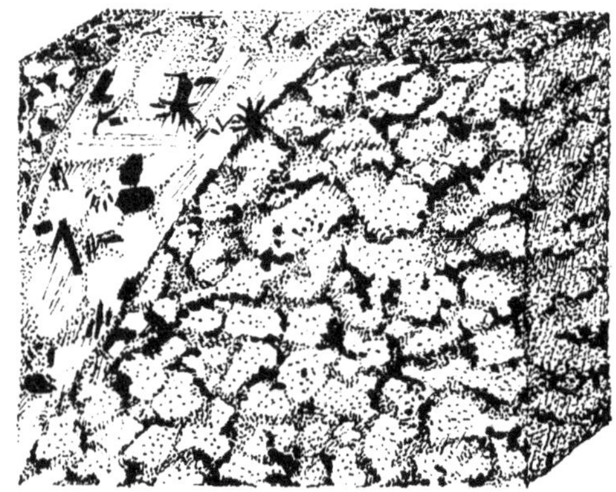

Granite slab with pegmatite veinlets (left half)

Volcanic and intrusive rocks that have been exposed by erosion at the Earth's surface are liable to alteration. Water, oxygen in the air, wind, radiation from the sun, and living organisms all contribute to the weathering and decay of rocks. Some minerals, such as garnet, zircon, magnetite and quartz are relatively stable, but other minerals readily alter. Some parts go into solution while other secondary minerals such as limonite, calcite and dolomite are formed.

The products of weathering are eventually carried away from their place of origin by water, wind and ice, and are deposited elsewhere. The weathered material so accumulated, regardless of whether it is compact or unconsolidated, is known as sedimentary rocks. Sedimentary rocks, which form only a small percentage of the Earth's crust, are of very considerable importance to man, notably as soil for growing crops. A well known sedimentary rock is sandstone. It is produced by the disintegration of pre-existing rocks, particularly granites, the fragments of which are carried and sorted by water and eventually deposited as sand consisting mainly of quartz, but commonly also with some feldspar. The grains are eventually cemented together by calcareous, siliceous or limonitic cements. Repeated changes in the current, which is normal with any flowing water, results in a visible layering in the sandstone which may be further emphasized by the presence of layers of different minerals.

Rock waste produced by the weathering of igneous rocks is, as stated above, transported mainly by water. However, at the same time some weathering products (mostly carbonates, sulphates and chlorides of the alkali, sodium and potassium, and alkaline earth metals, magnesium, calcium and strontium) dissolve and are responsible for the hardness of drinking water in some areas. These dissolved salts are eventually precipitated when the water evaporates, perhaps in some inland lake or sea. It is in this way that the sedimentary rocks known as evaporites are formed, as for example rock salt and calcareous sinter with its characteristic encrusted formations.

Not all dissolved materials are precipitated as salts (also called chemical sediments). Animals absorb a considerable amount to build up their tissues or skeletons (as in shells), and after they have died, this relatively insoluble matter may accumulate. These remains can form massive deposits of biogenic or organogenic sedimentary rocks. Examples of such sediments are some limestones, often rich in fossils, and diatomaceous earth.

Coal, too, is organic, although it is a sediment that has not been transported. Coal occurs in layers or seams that may be several metres thick, usually interstratified with shale or sandstone. It is formed predominantly from land plants

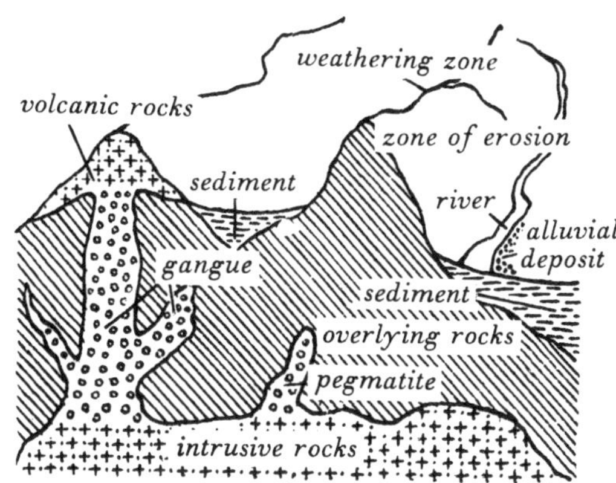

Formation of rocks

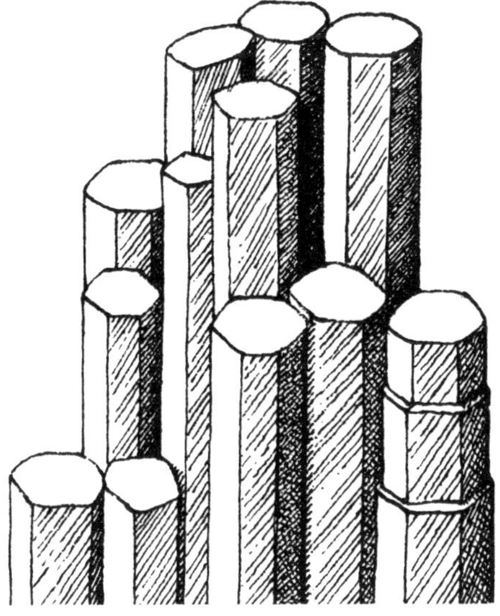

Basaltic columns (outline forms of the cross-sections varied)

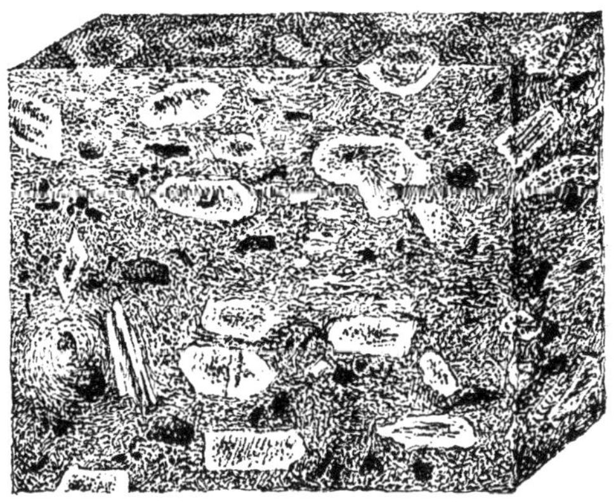

Quartz porphyry slab

through the process of carbonization (decomposition in dry conditions), which results in an increase in the carbon content, and a reduction in the hydrogen, oxygen and nitrogen content of the original material. The products of carbonization that are most important economically are peat, lignite, brown coal, hard coal and anthracite.

Metamorphic rocks comprise a major group of rocks which are genetically distinct from igneous and sedimentary rocks. Although they often differ only slightly in mineral composition from igneous rocks, in so far as minerals such as quartz, feldspar and hornblende are common in both rock groups, they are texturally distinct in that their constituent minerals usually have a preferred orientation. This is particularly striking in rocks in which minerals with a platy or tabular habit, such as mica or chlorite, predominate. This orientation of the minerals affects the physical properties of the rock, for it tends to split easily along the plane in which the minerals are orientated, and this is known as the foliation or schistosity.

It is comparatively easy to demonstrate by their field relationships and chemical compositions that metamorphic rocks are the product of the recrystallization of pre-existing rocks. The name 'metamorphic' (meaning change of form) derives from this. The grain-size of rocks tends to increase during recrystallization, which takes place in the deeper parts of the Earth's crust under conditions of elevated temperature and pressure, and affects large rock masses. The upward migration of magma and mountain building movements are usually connected with these processes, during which igneous and sedimentary rocks are metamorphosed, and acquire the foliation characteristic of the metamorphic rocks.

Gneiss is a common example of a metamorphic rock. Its essential constituents of feldspar, quartz and mica are the same as those of granite, but it differs in having a foliated texture and a layered or banded structure.

It can sometimes be difficult to identify rocks. It is essential first to identify the constituent minerals of the rock and these may sometimes be so small as to make identification difficult using the naked eye. Determination of hardness, streak, cleavage and crystal form are frequently impossible, because of the way in which the constituent minerals mutually interfere. In such cases rocks can be identified with certainty only under a microscope. For such an examination, which is essentially a determination of the optical properties of the minerals, thin sections are used and a brief word about their preparation is appropriate here. A fragment is chipped from the rock, or a slab a few millimetres thick is

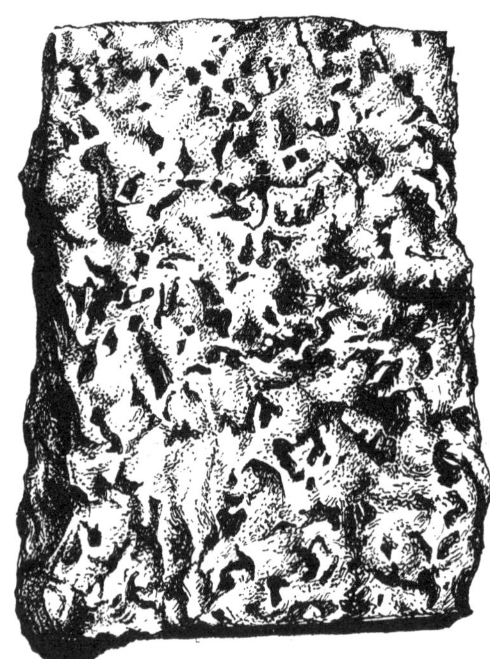

Nepheline-syenite

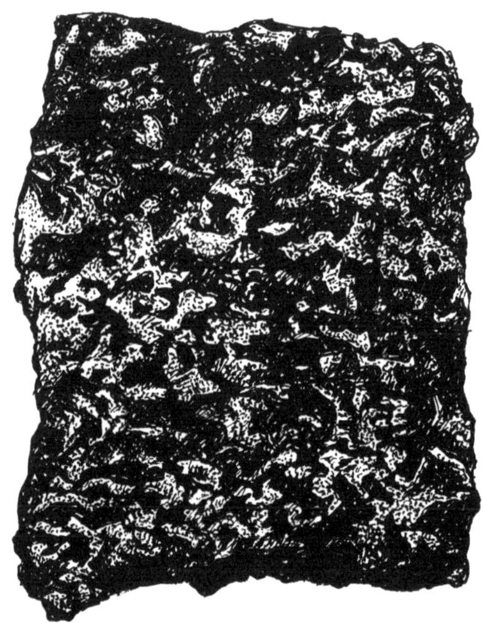

Diorite

cut from it with a diamond saw, which is preferable as it avoids crazing, which frequently occurs with chipping. The rock slab or fragment is then rubbed down until one side is smooth, using first coarse and then fine abrasive. Care should be taken to produce a very smooth flat surface. The polished slice should be about 3 cms by 4 cms in size, and it is cemented by Canada Balsam or other special adhesives to a glass slide (size: 28 mm by 48 mm) which has been roughened and warmed beforehand. The formation of bubbles, which would affect the optical measurements must be avoided. Carborundum powder of varying grades is then used to grind the specimen now firmly mounted on its glass slide, to a thickness of between 0.02 mm and 0.03 mm, when the majority of rock-forming minerals become transparent. The finished thin-section is then covered with a cover slip (size: 20 mm by 25 mm, thickness: 0.17 mm) which is again cemented with Canada Balsam. Canada Balsam is particularly suitable for mounting thin-sections because its refractive index is slightly less than that of quartz, and a little greater than that of orthoclase feldspar. Both minerals occur in rocks and can easily be identified in sections mounted in Canada Balsam.

Besides refraction, the determination of double refraction or birefringence is very helpful in mineral identifications. The refractive index varies according to the vibration directions of the light in the crystal. Doubly refracting minerals, when rotated through 360° between crossed polars on the stage of a polarizing microscope, become alternately light and dark four times. Moreover, when in the positions of maximum illumination, they show interference colours from which the birefringence can be determined. A range of other optical properties of the minerals in a thin-section may be determined using a polarizing microscope, which helps with their identification, and hence the naming of rock. However, it is not possible to give here further details of the optical properties of minerals, and reference should be made to the specialist literature.

Meteorites are in a category by themselves. The name is derived from the Greek and means appearance in the sky or air. Meteorites fall to Earth from space as solid objects. Large meteorites become so hot during their flight through the atmosphere that they may have a plainly recognizable fusion crust. To date, some 2,000 separate meteorite falls have been recorded. Scientific research began with a meteorite that fell in Zagreb, Jugoslavia in 1751. The size of individual meteorites varies enormously, ranging from dust particles to masses several metres across. The weight likewise varies from several hundredths of a gramme up to sixty tons. They are classified as irons, stony-irons or stones, according to their composition.

Eye-gneiss

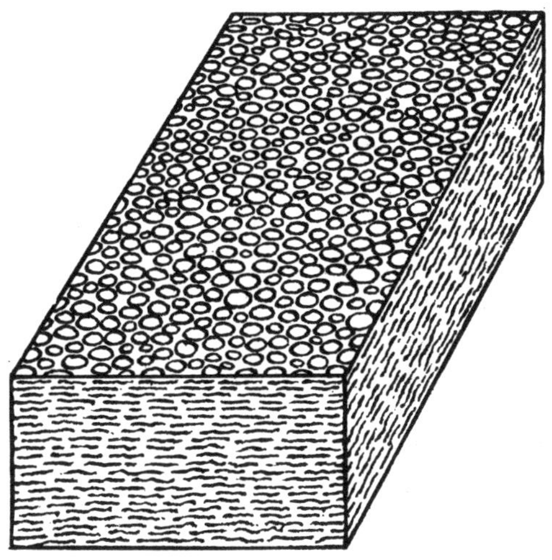

Slate-like texture

Iron meteorites consist almost entirely of nickel iron. They have an irregular shape, and the surface is usually indented, furrowed, or smoothly rounded. A surface that has been cut and polished, and then etched with acid, often shows a network of lamellae, up to one millimetre across, of the nickel compounds kamacite and taenite. This structure, which is characteristic of iron meteorites, is known as WIDMANSTÄTTEN structure, after its discoverer, A VON WIDMAN-STÄTTEN.

Stony meteorites, which make up the overwhelming majority of known falls, are recognizable by their black, and often shiny surface. Some stony meteorites contain small, round bodies called chondrules, composed of olivine, and are consequently known as chondrites. Other minerals besides olivine that are present in stony meteorites are plagioclase, feldspar and enstatite.

The third group of meteorites are the stony-irons, which include the rare pallasites. These are characterized by drop-shaped inclusions of olivine in a nickel-iron matrix.

The external appearance of tektites is very similar to that of meteorites. Their furrowed, grooved and corroded surfaces suggest a cosmic origin, but they are quite different from the meteorites described above. Tektites consist of black or bottle-green glass, and vary in shape from disc-like, to button-shaped, dumb-bell, spherical, tear-drop or pear-shaped. Their names derive from the regions where they were found, such as australites from Australia, philippinites from the Philippines and indochinites from S E Asia.

Tektite glass is composed predominantly of silica, but contains some aluminium, calcium and potassium. The origin of tektites has long been a subject of debate. The theory that they are the product of rock melted by the impact of meteorites with the Earth, is still the most widely accepted explanation.

Fossils

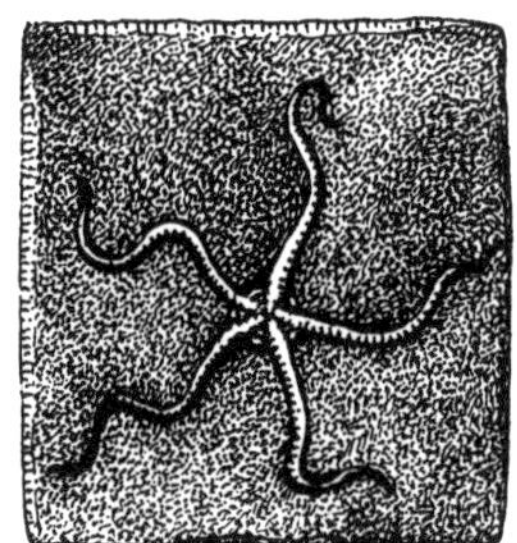

Left: platycrinites   *Right: palaeocoma*

Populus (Neo-cretaceous Period)

Fossils are the remains of prehistoric plants and animals, and their imprints (their footprints or tracks, for example) which have been preserved in sedimentary rocks. They provide reliable evidence about the development of organisms and the history of the Earth, and are characteristic of the various periods of the geological time scale. G Agricola extended Aristotle's original use of the word fossil to cover all objects that were dug up—whether fossils, as we now know them and previously known as petrifactions, or minerals. He derived the term from the Latin verb, *fodere*, meaning 'to dig' or 'to dig up.' Although the term fossil increasingly came to be accepted as referring to the petrified remains of prehistoric animals and plants, A G Werner and his students adhered to Agricola's original wider definition, and continued to include minerals along with fossils. The first separation of fossils from minerals is to be found in C Gesner's work, *De rerum fossilium*, which was published in 1558.

Fossils are abundant in sedimentary rocks, especially in limestones and shales, since animals living predominantly in the water are much more likely to be preserved when they die by becoming entombed in the sediment accumulating around them.

Fossils are formed under the most varied conditions. Since the soft parts, that is to say the organic substance of the living creature, and occasionally even the hard parts formed of calcium carbonate, silica, or chitin, decompose, only a small proportion of prehistoric animal and plant life has been preserved as fossils. Only rapid covering of the carcase by mud, sand or volcanic ash provided the conditions for fossilization. Embedding may have occurred at the site of the organism's natural habitat, or after transportation by water or wind.

The preservation of fossilized animals and plants occurs as a result of chemical changes. Thus parts of the animal or plant may become impregnated with water-soluble substances, such as silica or lime. Specimens that have been mineralized by pyrite, marcasite or, rarest of all, silver, are greatly prized by collectors.

Cavities within organisms such as shells or skulls may have been filled with mud or sand, which then formed an internal cast. Fossil remains in sand or coal for example, and also in chalk and limestone, sometimes served as a nucleus for the formation of concretions of water-soluble salts and superbly preserved fossils are often found inside such concretions.

Further, there are the exceptional examples of the preservation of creatures complete with soft parts by permafrost or salt-water, as for example fossil mammoths and rhinoceros. Finally, small organisms such as insects are sometimes found beautifully preserved in amber and other resins and these also are fossils.

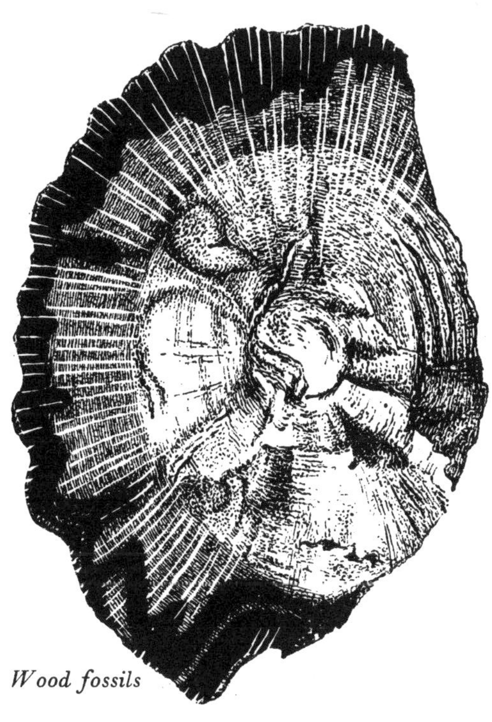

Wood fossils

Many fossils are of value in the determination of the relative ages of strata. However, it is essential for such fossils to have had a wide geographical distribution, and to have existed for only a fairly limited geological time span. These index fossils are of immense importance as time markers in tracing the course of evolution. Index fossils include some ammonites, trilobites, graptolites, brachiopods, bivalves, gastropods and some plants. For mineral collectors, silicified wood is of particular interest because silicification has preserved all the delicate detail of the structure of the plant. Silicified wood is generally so hard that it can be polished, which considerably enhances its appearance, especially when there are agate and fluorite inclusions.

Many fossil plants have been preserved as a result of carbonization. Club moss (*lycopodeum*), horse-tail, ferns, ginkgo and conifers are the main examples of this fossil group. In the wide range of animal fossils, fine specimens of any group may be both attractive and interesting.

Synthetic Minerals

In recent years the interest in minerals and the urge to collect them have grown enormously. Most collections are intended for display and demand thus concentrates on brightly coloured minerals. The number of crystals on a specimen may be small, but each should be several centimetres across, and they should be as near perfect as possible. A well crystallized cubic crystal of intense colour is an attractive object, but a transparent octahedron, attached to the matrix at one corner only, is perhaps of considerably greater interest.

Very few specimens meet these ideal requirements, and consequently they are rarities. The supply of specimens from working mines and quarries, as well as from abandoned, but still rich, workings is insufficient to satisfy the demand. This gap in the market can be bridged by the manufacture of synthetic crystals. By starting with the appropriate materials, the problem of colour can be overcome almost without difficulty. According to demand, brilliant blue copper sulphate, yellow potassium ferrocyanide, lemon-yellow potassium chromate or orange-red potassium dichromate can be produced to meet the requirements for display quality specimens. A further means of meeting the high demands for specific types of crystal is by directing the growth process of the salts being precipitated from aqueous solutions. It is possible to produce not only large crystals, which are always in demand, but also to control the quality of the crystals.

The amateur can produce unusual, beautiful synthetic crystals in a short time and with relatively simple materials, without the time-consuming need to acquire a lot of theoretical knowledge. Remarkable success can often be attained within a very short time, and beautiful 'specimens' produced more quickly than anticipated. Admittedly, there are only a few salts available that will crystallize as attractively coloured 'minerals,' and these will not altogether satisfy the collector's demands for good quality display specimens. Technical conditions set a limit to the growth of 'wonderful specimens,' and the beauty and lustre of the crystals may fade within a few days, as a result of inexpert handling. Insufficient humidity in dry-overheated rooms will reduce perfect, coloured crystals to powdery, worthless objects. Excessive humidity, on the other hand, causes some salts to dissolve and by the time that stains are noticed on the wooden display stands, it is too late to do anything about it.

Hitherto, it has been customary for deeply coloured artificial 'minerals' to be displayed as eye-catching exhibits alongside their poorly crystallized natural representatives. Vivid azure blue copper sulphate crystals were ideal for this purpose, and moreover they were easy to grow. Copper sulphate dissolves readily in water, and the simplest apparatus consists of a thoroughly cleaned large

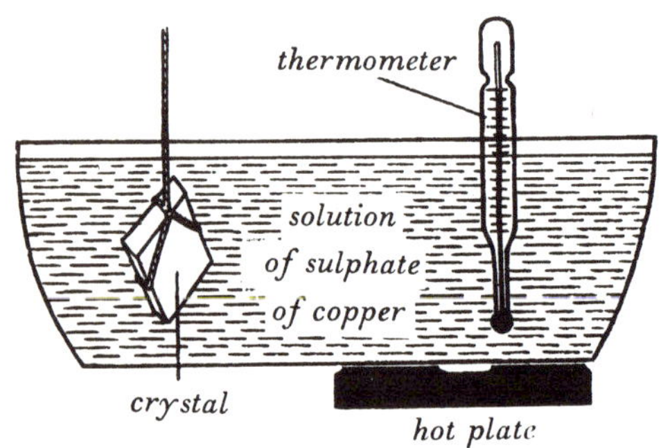

Methods of manufacture of crystals

beaker or glass filled with a saturated solution of copper sulphate which is prepared by heating water to between 40–45 °C, and adding copper sulphate until no more will dissolve. A cleaned piece of a copper sulphate crystal, preferably with the fewest possible exposed surfaces (usually fragments of broken crystals) is then placed in the glass, and the clear solution added. The glass is then covered to exclude dust, without preventing the evaporation of water, and stored away from vibration, and preferably in a room where the temperature is constant. After a few days fine blue crystals will also form on any scratches on the glass. At the end of the first phase of the experiment, the original specimen will be covered with a coating of small, radiating brilliant blue crystals, which is broken only at the smooth surfaces of the crystal chosen as the substrate. The tiny crystals serve as seeds for the continuation of the experiment. Their number can be reduced to just a few on the projecting parts, from which the large 'eye-catching' crystals will grow, and they are removed to grow further crystals from a solution prepared in the same way as before. After repeated transfers to new solutions, aggregates of crystals, a centimetre or so across have usually developed. Large, well-formed, single crystals can be grown by using the same method. Instead of placing a group of crystals in the solution a single crystal is suspended in it by a thread or a thin wire, to act as the seed crystal.

Single crystals can be grown more elegantly by constantly adding more of the nutritive solution. This can be done in a variety of ways. A large, oval dish containing the solution is heated by placing one end on a thermostatically controlled hot plate, so that a temperature gradient is constantly maintained with the crystal suspended in the solution at the other end. Repeated contact with saturated solution accelerates the growth. Two large glass tumblers or beakers connected by two, heat-insulated tubes can also be used. The salt solution is prepared in one vessel by gentle heating, and is driven by means of an agitator into the other vessel, in which the seed crystal is suspended. Scrupulously clean conditions and a working area free from vibration or movement are essential for the successful growth of crystals.

Occasionally weathering will cause successfully grown crystals to decay. The application of a colourless lacquer will help to preserve the crystals temporarily, although the vitreous lustre of the varnish reduces markedly their natural appearance.

The methods described here for copper sulphate can be adapted for other salts. The crucial factor is the solubility of the salts in water.

The vivid orange of potassium dichromate is just as popular as the blue

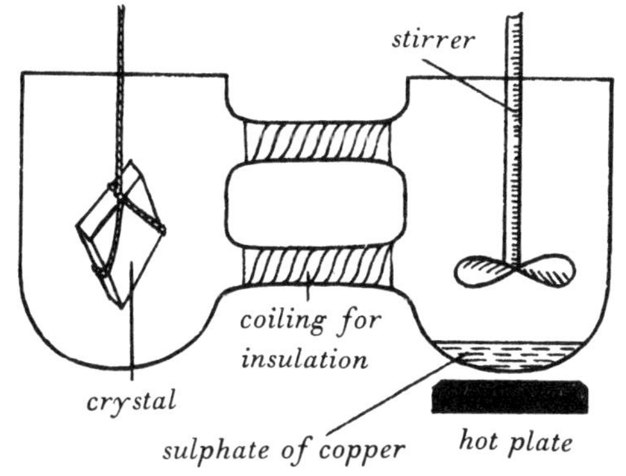

Methods of manufacture of crystals

copper salts. During the slow growth of large, tabular crystals, the salt becomes purified and this intensifies the brightness and strength of the colour. Yellow potassium ferrocyanide adds another colour to the range. Although its delicate colouring fits in well with the other 'artificial crystals,' it has never been popular. The crystals are rather soft, require extreme care in handling, and react strongly to changes in temperature.

Alum is a salt which allows one to determine not only the colour of the crystal, but also its intensity. By adding urea to the solution, large, single crystals can be grown, and these may be coloured pale red to purple by adding eosin, and sky-blue to deep cornflower-blue colours are produced with methylene-blue.

For the mineral collector, crystal growing as a hobby will necessarily be restricted to water-soluble salts. Growing crystals from aqueous solution is in reality a re-crystallization of the salt, and will always delight the lover of the beautiful brightly-coloured crystals, but without ever being of any commercial value. All other compounds, which crystallize from a melt, require—apart from costly apparatus—a fundamental and specialist knowledge and years of professional experience. These synthetic products are stable and are being used increasingly in industry.

The idea of producing minerals artificially is closely connected with the imitation of rare and costly gemstones. The widespread desire to demonstrate power and wealth in this way gave rise to the production of outright forgeries; a practice by no means confined to modern times, but one already known in the Middle Ages and in antiquity.

The earliest imitations were made of glass, for this was produced in ancient Egypt. Unlike synthetic stones, whose chemical and physical properties are virtually identical with real ones, it is mostly only the colour of glass imitations that resembles the real gems. It is worth remembering in this connection the discovery of J Strass, who melted together flint, alumina, chalk, iron oxide and soda to produce a glass, which it was possible to cut like a gemstone, and whose fire and brilliance came quite close to those of real diamonds.

The synthesis of ruby, perfected by A Verneuil in 1892 in Paris, and published in 1902, was an important industrial development. He fused alumina in a furnace and from the melt he managed to produce pear-shaped crystals. The addition of chromium oxide enabled him to produce red rubies. In order to achieve a successful and a perfect product, it was essential to use very pure starting materials. The Swiss watch-making industry showed great interest in this new development. Rubies are the jewels (bearings) used in watches, and precision is ensured by their hardness which reduces wear. The first plant for the manufacture of synthetic rubies was set up in Switzerland in 1914. Five years later a division of the chemical industry in Bitterfeld, began to manufacture the same product.

After ruby, there followed sapphire, which could be produced as a blue, yellow or green stone, according to which a particular metallic compound was added. The supply of synthetic gemstones now expanded continuously. Spinels and emeralds of convincing quality soon appeared. To date diamond is the only synthetic gemstone that cannot be produced in size as requested.

The hardness of synthetic gemstones has been exploited in drilling, polishing and cutting equipment. Advances in technology meant that the demand for pure, high quality crystals for optical instruments could no longer be met. Thus the synthesis of quartz crystals and a series of technologically important, inorganic compounds have been developed, and successfully brought into practical use. We would mention in conclusion the production of synthetic crystals for use as semi-conductors and in lasers. Information about their production and application can be found in technical journals.

Since earliest times collecting has been an avid pursuit of man—the objects collected and the motives for so doing, naturally changing with the ages. The regard for mineral raw materials was perhaps determined by the importance they had for everyday life. As the source of both precious and non-precious metals, mineral ores were valued just as much as precious and semi-precious stones. The colours and the peculiar shapes of minerals have always been impressive and a source of particular fascination. Magical powers ascribed to minerals were one of the reasons why they found their way into sagas and legends. Collecting is, however, also an ideal leisure pastime that increases one's knowledge and extends cultural horizons. The keen mineral collector is led to investigate the formation of minerals, to study the natural laws of the Earth's development, and as a result to become interested in mineralogy and geology. Through his collecting he may acquire valuable specimens that will also be of benefit to science. He will be preserving the past for the present and the future. Successful collecting, pursued with care and discrimination, has a formative effect on a person and contributes to the development of his personality.

The value of selective collecting became apparent, when, towards the end of the eighteenth century, existing and readily accessible mineral deposits close to the Earth's surface ran out. New mining methods had to be developed in order to meet the increased demand for raw materials. This led to the need to pass on precise information about the occurrence of minerals, their specific properties, the means of extraction and their uses. Hitherto information and experience had been passed on mainly by word of mouth, a system that was no longer adequate to meet the prevailing circumstances. Existing collections now provided a source of important, practical information as to geological and extractive conditions. They were welcomed by the newly-established schools of mines, were continuously expanded and were used increasingly for the training of mining specialists.

London— British Museum (Natural History)

One of the largest collections of minerals in the world is to be found in the British Museum (Natural History) in London. It was founded in 1753 and currently houses some 282,000 items. Of these, 180,000 are minerals, 95,000 rocks, and 2,000 specimens of minerals from ore deposits. Furthermore, the collection includes approximately 2,000 gemstones and 3,500 meteorites. The only collection to surpass it is that in the National Museum of Natural History in Washington (Smithsonian Institution) opened in 1912, with a total stock of 380,000 specimens (120,000 minerals, 200,000 rocks, 50,000 ores, 4,500 gemstones and 6,000 meteorites).

The historical development of the mineralogical collection of the British Museum reflects almost all the characteristics and problems common to many major national collections. In 1753 Parliament voted to purchase the best natural history collection of the day, a private collection belonging to Sir Hans Sloane, physician to Queen Anne. It was housed in Montagu House in Bloomsbury. Besides the notable herbarium, the 79,000 items included coins, medals, antiquities and a stock of 40,000 books and manuscripts. This collection was the first in the world that could be viewed for a fee. Yet there was no indication in those early days that the Museum was to acquire its present worldwide renown. Real scientific activity began at this institution only with the bequest of C Hatchett's collection of 7,000 minerals, and a further 836 by C M Cracherode. Inevitably, lack of space made it essential to move, and the transfer to the present building at South Kensington took place in 1881. Here the facilities necessary for the scientific development of the Museum were available. The collections were catalogued, and today the Museum is equipped to investigate minerals, meteorites and rocks using both classical and modern instrumental techniques.

The international importance of the Museum grew with the additions of further, valuable and wide-ranging collections of minerals, only a few of which can be mentioned here—1810; C F Greville (14,800 specimens), 1860; T Allan, R H and R F Greg (rich in type material), 1914–1938; F N Ashcroft (three collections with zeolites, especially from Ireland, and a collection of Swiss minerals), 1964; Sir Arthur Russell (14,000 minerals, mainly from Great Britain) and 1973; A C D Pain (gemstone collection from Burma). The mineralogical collection of the British Museum is regarded by mineralogists as the finest in the world, its prize items being minerals from Great Britain and Switzerland. The wide selection of sulphides, sulphosalts and zeolites is magnificently displayed and collections of calcite and fluorite are much admired. The Museum is one of the richest stores of newly discovered minerals and of type material. There are 8,000 specimens on view, representing about 1.5% of the total contents of the collection.

The development of France's largest collection followed much the same course as in Britain. The collection was founded at the École Nationale Supérieure des Mines (National School of Mines) in Paris in 1794. Its stock currently comprises 65,000 minerals, 15,000 rocks, 4,000 specimens from ore deposits, 700 gemstones and 300 meteorites. The collection was housed initially in the Hôtel Mouchy, the original home of the École des Mines. At that time its mineral showcases contained specimens from all over the world, but predominantly from the French Republic, which even then were arranged regionally. By decree of the Public Welfare Committee formed after the Revolution, the temporarily requisitioned private collections of GUETTARD, DIÉTRICH, JOUBERT and ALLAVOISIER were released, and placed under the care of the mineralogist, RJHAÜY. The collection expanded rapidly, meanwhile acquiring amongst others, a collection of 500 items from the Freiberg Mining Academy, which had been classified by AGWERNER. As a result of this expansion, in 1815 it was necessary to transfer the collection to the Hôtel Vendôme in Paris. In the course of time important private collections were added to the collection, including those of the MARQUIS DE DRÉE (after 1845, 15,000 items), ADAM (1881, with 7,000 minerals), DELESSERT (1888, 15,000 items), BERTRAND (1910, 2,000 specimens) and GLASSER (1949, several thousand specimens) to name only a few. Between 1957 and 1962 this, the largest collection in France, was completely rearranged. The underlying principle was to emphasize the beauty, rarity, classification and the uses of the minerals. The centre of attraction are the outstanding displays of specimens from European and African mineral deposits, while recent discoveries are also well represented. A great advantage in this respect are the close contacts maintained by the Museum with the laboratories and the mineralogical department of the Bureau de Recherches Géologiques et Minières (BRGM) in Orléans-La Source.

The decision to establish a mineralogical collection was taken earlier, by Peter the Great, in Russia than it was either in England or in France. In 1716 he ordered the purchase of a private collection, consisting of 1,195 specimens, from MDGOTTWALD of Danzig. This was the beginning of the mineralogical section of the Curio Cabinet of the Chamber of Treasures of St. Petersburg, which in 1725 was incorporated into the newly-founded Academy of Sciences. Twenty years later, MVLOMONOSOV published his first classification and the results of his mineralogical studies of the now expanded collection. Further additions to the collection of the Chamber of Treasures resulted from expeditions to Saxony,

Great Britain, the Urals, the Altai and the northern regions of Russia, undertaken by Academy members, P S Pallas, Y Y Georgy, Y Y Lepechin, S P Krasheninnikov. In 1780 a special exhibition of minerals was opened to the public, and to students, in particular. The collection reached a high point under the directorship of VM Severgin, who began scientific work on minerals, which had increased to 20,000 by 1836. Thereafter, mineralogy was pushed into the background by the upsurge of interest in palaeontology and geology, and this development was underlined when the name of the Museum was changed to the Geological Institute, which incorporated a mineralogical department. Fourteen years later, under V Y Vernadsky, the collection regained its independence, and became the centre of scientific research. After the October Revolution, the Museum's stock expanded considerably. The additions were the outcome of many expeditions led by A E Fersman, and the acquisition of excellent private collections.

In 1934, after Moscow had become the capital of the Soviet Union, the Academy of Sciences decided to transfer the collection from Leningrad to Moscow. The Museum was named in honour of A E Fersman in 1936 in recognition of his devotion to the care and growth of the collection. Currently the collection houses over 120,000 items, predominantly from ore deposits, and is classified according to various disciplines—meteorites (337), new discoveries from the Soviet Union, the geochemistry of the processes of mineral genesis, systematics, pseudomorphs, as well as gemstones and decorative stones.

Leningrad— Institute of Mining

Another important mineralogical collection in the Soviet Union is the one attached to the Mining Institute, which was founded in St Petersburg in 1773. Its chief features are the outstanding specimens from Russian ore deposits. The prize exhibits are a mass of deep green malachite, weighing 1,504 kg from a mine near Gumeshevo in the Urals, a gift of the Czarina Catharine II, a 500 kg specimen of smoky quartz and a beryl (24 cm long, weighing 2.5 kg), also from the rich mineral deposits of the Urals, as well as a specimen of native copper weighing 860 kg from the Republic of Kazakhstan. In 1825 a writ issued by Czar Alexander I suspended mining activities in order to permit specimens of noble metals to be extracted for transmission to these collections. As a result of this order the Institute of Mining of Leningrad came to possess a 36 kg nugget of gold, and beautiful platinum nuggets were likewise acquired at this time. The international character of the collection is emphasized by the purchase of the private collection of an English mineralogist, A Forster.

After the October Revolution, the development of the Leningrad collection was radically changed. In particular, deficiencies in the field of mining history and mining production were made good. With the inclusion of palaeontology, petrography, and the study of ore deposits, a comprehensive basis was provided for students training in the geological sciences.

Before the siege of Leningrad in the Second World War, the most valuable specimens were removed to Sverdlovsk, while the bulk of the collection, carefully packed, survived the worst years, sunk beneath a dam. Almost all the exhibition rooms which had been destroyed were restored in time for the 175th anniversary celebrations of the Institute in 1948. The collection is now restored to its former glory, and the preservation of its traditions continues. The resurgence of the mining industry has provided new and notable display specimens. One 800 kg specimen of smoky quartz and huge topazes weighing 10 kg from the Ukraine, and a blue fluorite crystal (330 kg) emphasize the tremendous development of the Institute.

Prague— National Museum

The mineral wealth of the Erzgebirge was the basis of the formation of important collections in Czechoslovakia and in Saxony. Centuries of mine-working yielded many excellent specimens of precious silver ores and of mineral species that occur together with tin and uranium. The acquisition of the extensive private collection of COUNT K STERNBERG of Březina by the Národni Museum (National Museum) in Prague in 1818 marks the beginning of this collection, which now enjoys world-wide esteem. Bequests and the dedicated work of the curator, F X M ZIPPE, led to the rapid expansion of the Museum's stock, the chief sources of the material being the ore deposits in Bohemia, Moravia and Slovakia. Exchanges yielded important acquisitions from all over the world, so that by 1824 it presented to the public one of the most comprehensive and scientifically organized collections. After the removal to the building in Wenceslaus Square had been completed in 1895, K VRBA began to build up specialized collections of precious and semi-precious stones, as well as meteorites and tektites. The close collaboration between the National Museum's scientists (J KREJCI, E BORICKY, L SLAVIKOVÁ) and those of the University of Prague (F SLAVIK, J KRATACHVIL, C RUŽIČKA) resulted both in a high standard of display and successful research. The opening up of new deposits required documentation, and provided new material for the collection. Besides mineralogy, a large proportion of both the displays and the stock is taken up with petrography and the study of ore deposits. This is plainly reflected in the fact that the Museum's collection currently holds 120,000 minerals, 25,000 rocks, 13,000 gemstones, 400 meteorites, and 15,000 tektites. The outstanding exhibits of the Prague collection are a magnificent specimen of millerite from the coalfield in the Kladno basin, a cavity containing ilvaite considered to be unique, and a giant reniform mass of uraninite from Jáchymov. The large collection of moldavites (tektites) from southern Bohemia is without equal.

Berlin—Museum of Natural Science of the Humboldt University

In 1770, C A Gerhard founded a Mining Academy in Berlin, at which he also lectured in mineralogy and mining. In 1781 he sold his private collection to the Academy in exchange for a life annuity. His successor, D L G Karsten enlarged the collection appreciably, and also catalogued it; his classification of the Museum stock was an outstanding achievement. Subsequently the collection has been rehoused several times. From the Academy of Arts it went to the *Jägerhof*, from there to the *Alte Münze* in the Werderscher Markt, but it was not until 1888 that it was incorporated into the *Museum für Naturkunde* (Museum of Natural Science) in the Invalidenstrasse. Its conversion into a mineralogical museum at the University founded by W von Humboldt in 1814, was crucial to the development of the collection. All the professors concerned with mineralogy, as a result of their special fields of interest, also determined the direction of the work of the Museum, and each of them endeavoured by means of gifts, purchases or exchanges, to extend its stock, and thereby to create a solid foundation for study and research. The collection was severely affected during World War II, losses were suffered while it was in store, and damage to the exhibition and store rooms was extensive.

Dresden—State Museum of Mineralogy and Geology

The historical development of the collection of the *Staatliches Museum für Mineralogie und Geologie* (State Museum of Mineralogy and Geology) at Dresden differs from that of the Freiberg and Berlin collections. Its origins go back to the Dresden *Kunstkammer* (Chamber of Art) whose existence is first recorded in 1560, and greatly benefitted by Augustus, elector of Saxony. The basis of the collection was not minerals, but rocks (marble, basalt, serpentine) from Saxony, which were collected for practical purposes by G M Nosseni. During the Thirty Years' War, expansion of the collection came to a halt, and the various changes in accommodation did not help either. Up to the time of the outbreak of the Seven Years' War, which also caused losses, the collection had been separated from the *Kunstkammer* to form an independent mineral collection. Considerable additions were made to the stock of the collection by the purchase of large private collections, which helped to stimulate scientific study. When, in 1847, H B Geinitz took over the directorship, the mineralogical museum underwent a rapid expansion. Although its specific field of study was palaeontology, mineral collecting was not neglected. The reconstruction of the exhibition as well as the extension of its stock are the outstanding hallmarks of his fifty-one years as director. Mineralogy once again took precedence under E Kalkovsky, who was especially concerned with enlarging the collection with the guidance of RJ Baldauf, the keenest of collectors. In 1940, the famous Baldauf collection, comprising almost 10,000 items, was purchased for the sum of 250,000 Reichsmark. Wartime conditions prevented it from being mounted for display, as the entire work of the Museum was directed towards the evacuation of its contents. The war destroyed the exhibition rooms as well as a large proportion of the catalogued material, but the mineralogical collection was least affected. Reconstruction work began in 1946, with the return of the stock that had been sent away for safe-keeping, and currently it comprises 50,000 minerals, 15,000 rock specimens and 350,000 fossils.

Freiberg— Collections in the Geological Sciences Section of the Mining Academy

Right from the very beginning of its history, collections were of particular importance to the Freiberg Mining Academy. Their display was included from the outset in the plans for the foundation and equipment of the mining school. The 'Stuffencabinett' (Cabinet of mineralogical specimens) of that time was the nucleus of the time-honoured and world-famous collection of today. Its development has always been closely bound up with the scientific training of the students. A collection of mineralogical and geological specimens was already in existence when the Mining Academy was opened in 1766, (probably dating back to C E Gellert) and was enlarged by donations from both founders. By means of purchase and exchange, excellent mineral specimens reached Freiberg. With the acquisition of A G Werner's private collection in 1814 for the sum of 40,000 Taler, the collection attained its highest level. During his forty-two years as Professor of Mineralogy, A G Werner had built up his collection to a total amounting to almost 12,000 specimens, and had used them constantly for research and in teaching. It thus reflects the position of the geological sciences of that period, to which his work had given considerable impetus. The setting up of separate geology and mineralogy lectures in 1816 brought in its wake a partition of the collection, which was divided among the various professorships. A G Werner's collection, however, remained unaffected, and is to this day preserved intact as a historical entity.

A G WERNER's successor, C F C MOHS after whom the scratch-hardness scale was named—and F A BREITHAUPT, introduced the Linnaean system into mineralogy. Thereafter, every mineral was given a Greek or Latin generic and specific name. By the end of his term of office in 1866, F A BREITHAUPT had enlarged the resources of the collection to very nearly 20,000 specimens, to which his successor, A J WEISBACH added a further 10,000. The latter set himself the task of returning once more to the use of the common nomenclature for minerals. This required entirely new captions for every single specimen, and at the same time he classified the minerals, and re-arranged the collection. Notable additions were contributed by former students of the Mining Academy. They donated comprehensive collections accumulated in the course of their prospecting work, mostly from abroad. The rooms in the von Oppel Haus in the Futtergasse ultimately became overcrowded, and so under the directorship of F W L KOLBECK the collection was transferred unharmed to the newly erected A G WERNER Building in the Brennhausgasse, which was dedicated on the 21 May 1916. In the course of six years, F W L KOLBECK re-organized the collection according to the classification set out in *Synopsis Mineralogica* by A J WEISBACH and displayed it in virtually dust-proof 'Kühnscherf' glass showcases. The specimens were exhibited behind large panes of special glass in five ascending rows, so that having been carefully selected according to shape, size and colour, they were displayed to maximum effect. The format of the exhibition remains functional and effective today, making it a top-ranking example of its kind, even by world standards. Any gaps in the collection were assiduously filled, and the industrial expansion in Germany, with the opening of new ore deposits, was a significant factor in achieving this aim.

The outbreak of the Second World War brought work virtually to a standstill. The exhibits were moved to safety with relatively little damage, and by June 1945 reconstruction work had already begun, the arrangement of the collection following the former system, which was retained until 1968. During these years priority was given to work aimed at a complete reorganization of the systematic display according to modern principles that reflect current developments in science, while observing visual and aesthetic considerations. The new system—a combination of the classifications of H STRUNZ and I KOSTOV— inevitably improved the impact of the exhibits, while preserving earlier traditions, with a minimum of alteration to the stock of display specimens. The modification of the collection was completed in time for the beginning of the 23rd International Geological Congress in Prague in 1968. In the course of the preparatory phase a wide-ranging international exchange system was established, and generous purchases were made by the university.

The Freiberg collection has at its disposal many fine, exquisite display specimens from classic sites both in Germany and abroad, and is arranged from a paragenetic point of view.

Further large mineralogical collections are to be found in the capitals, and in the major cities of countries known for their mineral wealth. Of these, we shall just mention the following few: the Naturhistorisches Museum in Vienna (130,000 items), the Mineralogical Geological Museum in Oslo (210,000 items), the Royal Ontario Museum in Toronto, Canada (162,000 items) and the Geological Department of the University of Alberta in Edmonton, Canada with 230,000 items.

Hydrothermal Minerals

Hydrothermal ore deposits are formed by deposition from hot to warm solutions. The minerals are deposited mostly in veins or joint cavities into which the solutions penetrate. Sometimes the veins are heavily concentrated in a relatively small area and so may produce an ore deposit. Ore veins can be categorized according to the predominant mineral present. Hydrothermal deposits characteristically contain a large number of minerals, and these commonly fall into distinct groups, which are often deposited in a particular sequence.

Minerals of Lead-Zinc Deposits

Mineral deposits of lead-zinc type are relatively common in the British Isles, and at some localities have been worked since the Middle Ages. For instance, ores of lead and zinc have been worked in the Lake District since the sixteenth century and flourished in the 1850s. Similar ores occur in the Isle of Man and cover an extensive part of North Wales, including Anglesey, the Dolgelley area, Snowdon and Llanrwst. In all these deposits the ore occurs as veins and is associated with volcanic rocks. Lead and zinc have also been mined in central Wales and the Welsh borders, and galena was produced from vein deposits in Northern Ireland in an area east of Belfast.

Page 47
Galena with siderite and quartz (crystal combination forms: of cube, octahedron, and icositetrahedron)
Neudorf, Harz (G.D.R.)
Size: 13 cm × 9 cm × 5 cm

Galena in fluid form
Zug near Freiberg (G.D.R.)
Size: 4.5 cm × 6.5 cm

Extensive metalliferous deposits of lead
and zinc occur in the Southern Uplands of
Scotland, notably in the Leadhills-Wanlockhead district, where galena was mined
from the seventeenth century until the 1930s
and again in 1958. Over 70 veins were known
in the Leadhills-Wanlockhead area. Lead
ores have been mined in the Pennines since
Roman times. Galena is the principle mineral
and it has an average silver content of about
200 parts per million. Other important minerals are sphalerite, chalcopyrite, chalcocite,
and covellite. Extensive fluorite deposits are
associated with this mineralization and are
still mined.

Although ores of tin and copper are the
most important of those found in the mining
districts of Devon and Cornwall, lead and
zinc have also been produced on a significant
scale. Many of the lead ores, principally
galena have proved to be relatively rich in
silver. For instance the galena mined near
Combe Martin, on the north Devon coast,
had a reported silver content of 0.51 %.

In the near-surface oxidized zone of lead-
zinc ore deposits, introduction of oxygen and
often water can lead to the formation of
crocoite, pyromorphite, wulfenite, mimetite
and mottramite.

*Galena with dolomite and chalcopyrite
(crystal forms: cubic and combination of
cube and octahedron, octahedral)
St Joseph Lead District, Missouri (U.S.A.)
Size: 14 cm × 6 cm × 12 cm*

Galena (Lead Glance)

This is the commonest lead ore and it is characterized by the shining silvery lustre of freshly fractured surfaces. Its Latin name— *galena*—also refers to its lustre. The heavy metal lead is the main constituent of the ore, giving the mineral its considerable density. This property of galena, and the perfect cubic cleavage, which appears as hairline cracks on surfaces which frequently have a brilliant metallic lustre, are important in the identification of the mineral. The cube is the basic crystal form of galena. Cleavage fragments, no matter how small, are also cubes or platy forms. Octahedra occur in only a few ore deposits. The number of faces on galena crystals increases by combination of the cube and other crystal forms. Galena also occurs massive and does not cleave into cubes but has a fine-grained or scaly fracture.

Galena has been found at many localities in the British Isles, and has been mined at a number of them. The principle areas are Derbyshire, where it is associated with the carboniferous limestone, and in Durham and North Wales. It is also found in the mineralized areas associated with the granites in the south west of England, and in Cumberland. Fine specimens also come from the states of Missouri and Illinois in the United States of America. The largest crystal (edge length of 25 cm) and the heaviest (30 kg) was found at Joplin, Missouri.

The extraction of silver has always been an important aspect of the smelting of galena. The silver content of the ore may be significant making it not only the most important of the lead ores, but also an important silver ore. The silver occurs as very small grains of silver minerals such as argentite, proustite and tetrahedrite which are intergrown with the galena.

Sphalerite with calcite and brown spar
Brand near Freiberg (G.D.R.)
Size: 12 cm × 12 cm × 5 cm

Sphalerite (Zinc Blende, Blende, Black Jack)

Although brass, a copper zinc alloy, was manufactured in antiquity, the extraction of zinc from its ores was however unknown. Sphalerite, the most important ore of zinc, was considered a barren lead ore. In the late Middle Ages the miners of Saxony and the Harz region called zinc blende, which occurs in association with galena, simply blende. The smelting of zinc blende presents difficulties, since metal is not directly produced, and even the separation of the other minerals present in the ore was a laborious operation.

It was not until 1734 that zinc, the main component of blende, was discovered by the Swedish chemist, G BRANDT. Although the name lost its original meaning, it was nevertheless retained.

Despite its softness and easy cleavability, clear crystals have been cut as gemstones. It commonly develops good crystal shapes, though they are rather variable in habit.

The name, sphalerite, derives from the Greek word *sphaleros*, meaning 'treacherous,' because it is sometimes difficult to distinguish it from certain other minerals.

The colour of zinc blende depends very largely on what other metals are present. When pure, it is colourless and is known as cleiophane, but the smallest amounts of impurities produce more typical yellow to brown crystals. The presence of iron produces darker tones ranging from red to brown and black. Ruby red zinc blende is particularly striking and particularly fine specimens come from Oklahoma, Kansas and Missouri in the United States of America. Compared with these, the iron-rich (up to 20 %) black zinc blendes known as christophite, named after the St Christoph mine in Breitenbrunn in the Erzgebirge in the German Democratic Republic, and marmatite (after Marmato in Peru) both look very insignificant.

Sphalerite (zinc blende) with galena and chalcopyrite
Banská Štiavnica (Czechoslovakia)
Detail

Black zinc blende from Freiberg is of historical interest. It was used by H T RICHTER and F REICH in 1863 to prove the existence of the element, indium. Zinc blende can also contain a number of other elements such as silver, cadmium, gallium, mercury, copper and tin.

The crystals are often tetrahedral forms, and less commonly rhombdodecahedra or cubes. Twinned crystals are often distorted and have curved faces.

Very fine specimens showing superb black crystals with an adamantine lustre come from Serbia in Jugoslavia and Tennessee in the United States of America. Other localities from which excellent crystals have been obtained are Alston Moor, Cumberland; numerous localities in Cornwall; Laxey in the Isle of Man; Weardale in Durham and from Wanlockhead and Dumfries in Scotland.

Wurtzite

Zinc blende and wurtzite have the same chemical composition. Both minerals are polymorphs of zinc sulphide, ZnS. Zinc blende is cubic, whereas wurtzite is hexagonal.

Wurtzite mostly occurs massive with a radiating, scaly or encrusted structure. The colour ranges between light brown and brownish-black. Zinc blende and wurtzite often occur in association.

The best known occurrences are at Stolberg, near Aachen (Federal Republic of Germany), Pribram in Czechoslovakia, Bytom and Chranzow in Poland, and Oruro in Bolivia. A massive fibrous form in which it is mixed with sphalerite is found at Liskeard and other Cornish localities.

Pyrite (Iron Pyrites)

Pyrite, which emits sparks when struck, has been used since ancient times. Different minerals were used to generate sparks for making fire, and were distinguished by their colour.

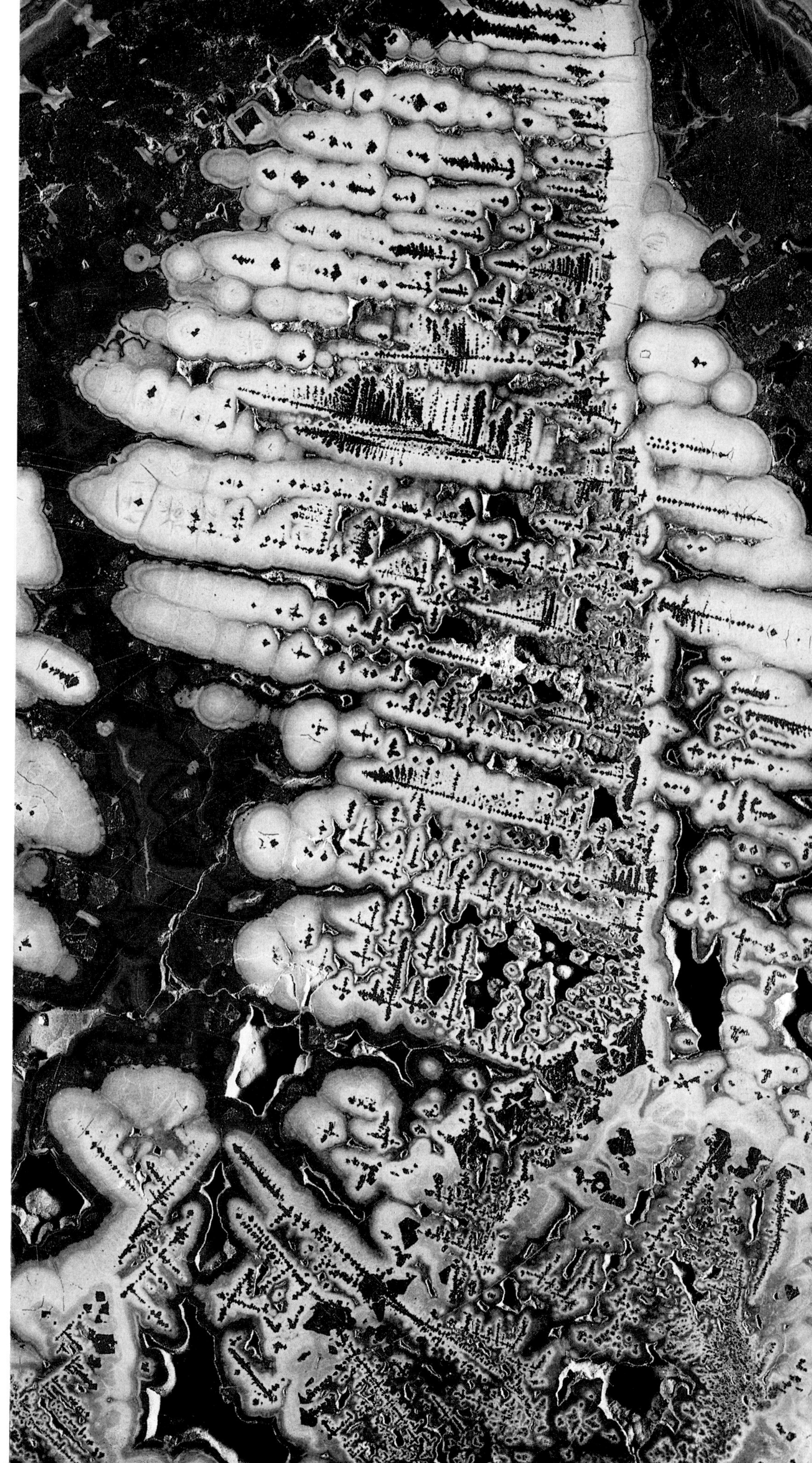

Sphalerite and wurtzite with galena
Aachen (Federal Republic of Germany)
Detail

These included the silver-coloured arsenopyrite and the gold-coloured chalcopyrite
(copper pyrite). The distinguishing feature
common to all these minerals is their light
colour.

The metallic yellow of pyrite, modified
only rarely by iridescent darker shades of
brown and green, makes it a highly-prized
collector's piece, while the rich, brassy yellow of chalcopyrite comes a very close second.
Distinguishing between the minerals is
easy—pyrite is hard and will scratch glass,
and it commonly forms cubes, while chalcopyrite is softer—can be scratched with a
penknife—and does not form cubes but
tetragonal crystals.

Pyrite crystals are usually cubes but pentagonal dodecahedra, also known as pyrito
hedra, and occasionally octahedra also occur.
An important diagnostic feature is the presence of striations on the cube faces, running
parallel to one pair of edges. They are perpendicular to each other on adjacent faces,
and are caused by the alternating growth of
cube and octahedron. Over sixty crystal
habits of pyrite are known as a result of
combinations of various forms.

Since pyrite is so abundantly and widely
distributed, it is possible to single out only a
few localities from which particularly fine
crystals have been obtained. These include
Liskeard, St Just, St Ives and many other
localities in Cornwall, and Tavistock in
Devon. Many of the old mining areas throughout the British Isles were a source of pyrite.
Fine material also derives from Rio Marina
on the island of Elba, Brosso and Traversella
in the Piemont (Italy), Ambas Aguas (Spain),
Rodna and Zlatna in Rumania, Kassandra on
Khalkidhiki (Greece), Murgul and Artvin in
Turkey, Quiruvilca in Peru, as well as Central
City in Colorado and French Creeks in Penn-
sylvania, United States of America.

Pyrite oxidizes readily to iron sulphate
thereby reducing its splendent lustre. This
will form sulphuric acid which destroys the
specimen and produces the dirty brown of
limonite.

Marcasite
Pitcher, Oklahoma (U.S.A.)
Detail

55

Marcasite

Until the beginning of the nineteenth century, marcasite was thought to be pyrite. However, it is a less stable compound than pyrite and is of much more limited occurrence.

Pyrite and marcasite are polymorphs of iron sulphide, FeS_2, but whereas pyrite belongs to the cubic crystal system, marcasite is orthorhombic.

Marcasite crystals are mostly tabular or spear-like. Twinning produces 'cockscomb aggregates.' Reniform masses, known occasionally as hepatic pyrite, and concretions ranging in form from coarsely radiating to finely fibrous are widely distributed. Marcasite differs little from pyrite in colour. Depending on the area and the conditions under which the minerals formed, greenish-grey tones can occur. Marcasite more commonly shows iridescence than does pyrite.

Especially beautiful, brilliant, spear-like crystals have been found in the open lignite workings in north-western Bohemia (Czechoslovakia). Tavistock in Devon is the source of cockscomb masses, and it also occurs at Weardale in Durham. It is also found in concretionary masses in the Chalk of southeast England. In the United States of America fine specimens are found at Galena, Illinois and at Mineral Point, Iowa County, Wisconsin.

Arsenopyrite (Mispickel)

Arsenopyrite derives its name from its properties and its appearance. As early as the sixteenth century it was known pejoratively as mispickel probably in reference to the repulsive smell of garlic the ore emits when struck.

It has a high arsenic content of 46 % (formula FeAsS) so that it was used to produce arsenic and the pigment orpiment. Its presence was unwelcome in the refining of lead and silver ores, since arsenic prevents the formation of slag, the silver is ruined and much of the lead is 'eaten.' Prior to smelting,

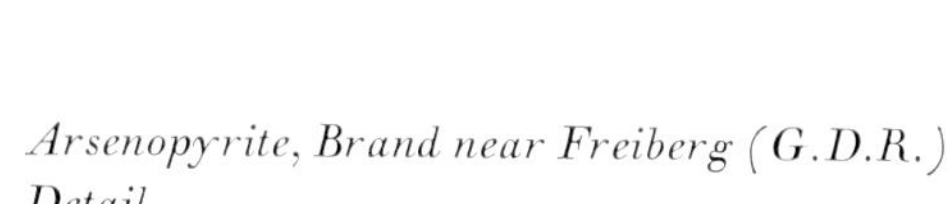

Arsenopyrite, Brand near Freiberg (G.D.R.)
Detail

56

it was essential to separate on the sorting table the sulphide ore from the harmful arsenopyrite and zinc blende.

Its pale, silver to tin-white colour and its hardness are similar to those of the commonly associated minerals arsenopyrite and pyrite. Columnar arsenopyrite crystals resemble those of marcasite. Arsenopyrite may contain small quantities of silver, gold, cobalt and nickel, all of which metals have a high commercial value. Large crystals have been obtained from the Tavistock area of Devon, and it has been found in many of the ore deposits of Cornwall. The largest arsenopyrite deposit in the world is at Boliden in northern Sweden. Other well-known localities are Mitterberg in Austria, the Trepca mine near Kosovska-Mitrovica in Serbia (Jugoslavia) and La Villeder in France.

Chalcopyrite (Copper Pyrites)

Until the Middle Ages chalcopyrite was considered of but medium importance by miners. It was a typical associate of minerals such as pyrite and arsenopyrite, and like them is a sulphide and was considered by the miners to have similar properties although it was distinguishable by its golden colour. However, copper could be smelted from this variety, and it was consequently known as an ore. The term chalcopyrite derives from the Greek words, *chalkos* meaning copper and *pyr* meaning fire.

Chalcopyrite and pyrite are very similar, but their different properties make it possible readily to distinguish between them. Chalcopyrite is not sufficiently hard to scratch glass, and it rarely forms good crystals; when it does they are of tetrahedral form as opposed to the cubes and pyritohedra of pyrite. Chalcopyrite is more readily identified by its striking and characteristic violet, greenish blue to greyish black iridescent tarnish. The specimen most sought-after by collectors has a light sprinkling of malachite or azurite, resulting from weathering of the chalcopyrite in the presence of carbonic acid.

Arsenopyrite with quartz
Weitzschen near Meissen (G.D.R.)
Detail

The colour of such specimens ranges from the
brassy yellow of chalcopyrite, the grass-green
of malachite to the inky-blue of azurite, inter-
spersed with clear or milky quartz crystals.

Deposits of chalcopyrite occur all over the
world, some of the localities that have yielded
good crystals being Laxey, in the Isle of Man,
St Agnes and many other localities in Cornwall; at Falun in Sweden it occurs as large
masses; also from many localities in France,
and from French Creeks, Pennsylvania in
the United States of America.

Silver

Silver was already well known in ancient
times, and its lustre gave it its name. Although native silver is less common than
gold, more is produced each year because
there are thirty-two minerals from which
silver can be extracted, as opposed to only
six gold minerals. The most important silver
ore is galena, with an average silver content
of 0.2–2 %. The classic relative worth of
silver to gold was established by a French
edict in 1785, according to which the value
of gold is 15.5 times higher than the same
weight of silver.

The rapid formation of grey or black tarnish has an adverse effect on the value of the
silver. This thin coating of silver sulphide,
that is hard to dissolve, coats the surface of
the metallic silver, and can be removed by
reducing agents (tin) in alkaline solution
(warm sodium carbonate).

*Pyrite with twin formation of the 'Iron
Cross'*
Transylvania (Rumania)
Size: 4 cm × 5 cm × 3 cm

Pages 60/61
Pyrite
Sparta, Illinois (U.S.A.)
Size: 8 cm × 8 cm × 0.5 cm

Pyrite
Quiruvilca (Peru)
Size: 10 cm × 11 cm × 4 cm

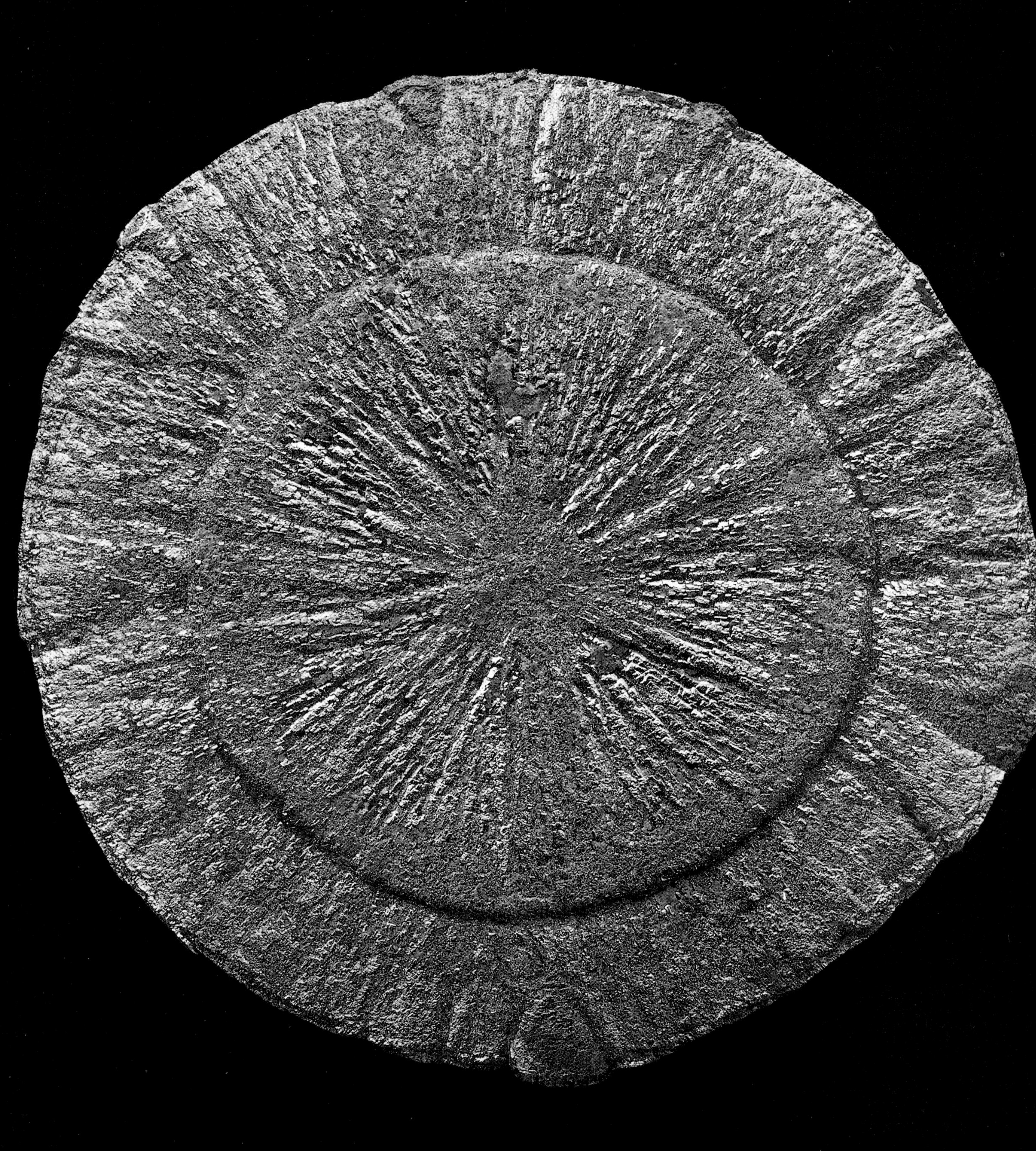

Silver occurs in many forms. Dendritic silver with individual masses sometimes reaching 30–40 cms in length; wiry and feathery forms, lamellae, and foliated masses are very common, and cubic crystals are also found. Intertwined bundles of silver resembling locks of hair and intergrown with jasper is a rare form of silver agate found at Johanngeorgenstadt in the German Democratic Republic.

The centres of silver mining in Saxony used to be Freiberg and Schneeberg. The Himmelsfürst mine in the Freiberg field was famous. In 1857 a silver specimen weighing 225 kg was extracted and contemporary records mention silver plates weighing up to 34 kg.

Evidence of this prosperous period lies in the churches and merchants' houses that survive in a splendid state of preservation in Freiberg, Schneeberg and Annaberg in the German Democratic Republic. Most of the silver was used for minting coins and medallions, for example the 'Schreckenberger' or 'Engels Groschen' of Annaberg. It was minted as early as 1498 from silver from the mountain of the same name on the western edge of the town. The two-headed 'Guldengroschen' pieces (1500) of Freiberg and Zwickau are also famous. The rich silver finds of Joachimsthal (Jáchymov) in Bohemia (Czechoslovakia) also described by G AGRICOLA in his *Bermannus* were the source of the large number of 'Joachimstaler' or 'Schlicktaler' that were minted in 1518. Not only was the currency based on this silver, but also considerable amounts of it were used for the manufacture of bowls, goblets, beakers, candlesticks and elegant filigree work. Today the greatest consumer of silver is the chemical industry, mainly for the production of photographic film, followed by laboratory chemistry with its large requirements for silver apparatus and silver salts.

Pyrite on hematite
Rio Marina, Elba Is. (Italy)
Size: 8 cm × 10 cm × 8 cm

Pyrite on hematite (combination of pentagonal dodecahedron and diakisdodecahedron, also known as diploid, and octahedron)
Size: 4 cm × 5 cm × 6 cm

Chalcopyrite
Grosschirma near Freiberg (G.D.R.)
Size: 7cm × 5.5cm × 2.5cm

Pyrite
Gavorrano, Tuscany (Italy)
Size: 20cm × 14cm × 18cm

Wurtzite
Brand near Freiberg (G.D.R.)
Size: 5cm × 5cm × 4cm

Perhaps the most famous silver mines are those at Kongsberg in southern Norway. Some of the magnificent specimens of native silver have been of huge size. These mines were worked continuously for more than 300 years from 1623 to 1957, during which time it is estimated that some 1,300 tons of silver were produced. In all there were about 130 mines being worked, the deepest of which, the Kongens mine, reached 1,076 metres. The silver content of the ore varied between 100 and 350 grams per ton, so that the profitability of the mines fluctuated widely. Some of the ore veins contained exceptionally high silver values and lumps of native silver weighing up to 500 kilograms were sometimes found. Large cubic crystals sometimes occurred with edges up to 17 cm long.

Many of the finest specimens from Kongsberg take the form of wiry masses, and these are to be seen in museums around the world. Particularly fine specimens can be seen at the Mine Museum. Cobalt and nickel minerals, including rammelsbergite and chloanthite, are present as coatings on many native silver specimens.

The principle minerals found in association with silver are quartz, fluorite, apatite, calcite, baryte, argentite and coal blende—a carbonaceous material containing more than 90 % carbon.

Outside Europe, the largest silver producers are Mexico and the United States of America. There are five hundred known Mexican mining localities with three thousand mines along the western slopes of the Cordilleras.

Argentite-Acanthite (Silver Glance)

The high yields of argentite and native silver were largely responsible for the silver wealth of Saxony. In his description of argentite, G AGRICOLA referred to its similarities to galena. In the Middle Ages the miners called this highly-prized silver ore 'glass ore,' since it can be cut with a knife, and the resulting cut surface is so smooth, even and shiny that it was compared with the surface of glass. When freshly mined, it resembles galena, but within a very short time it blackens. Because of its lustre, it has been called silver glance.

Both in colour and crystal form, argentite is similar to galena. Cubes, octahedra and rhombdodecahedra predominate; but it occurs also as distorted forms resembling teeth and wires, the latter often twisted along their length; capillary, reticulate and arborescent forms are less common. The last of these are especially interesting when they are intergrown with pale dolomite.

The highest yield of crystallized argentite came from the Freiberg field. A find made at Johanngeorgenstadt in the seventeenth century was quite exceptional both in size and purity. Pieces weighing one kilo were used to provide the silver used to strike medallions and for chased work, mainly with religious motifs.

Finds of almost equal value, but more varied in formation, are the magnificent specimens from Banská Štiavnica in Czechoslovakia. Last, but not least, are the deposits of argentite in Mexico, the country with the greatest silver wealth in the world. Notable finds of argentite associated with silver came from the ore veins of Zacatecas near Guanajuato. Argentite is found associated with the copper ores at Butte, Montana, in the United States of America and it occurs in large amounts in the Comstock Lode in Nevada.

An entirely different form of argentite was found in Joachimsthal (Jáchymov) in Bohemia in the middle of the nineteenth century. The pointed, thorn-like crystals give the specimens a spiny appearance. This new variety of argentite was named acanthite. (In fact the two are polymorphs—cubic argentite is stable only above 180 °C, below this temperature it has an orthorhombic structure and is called acanthite.)

Silver on calcite
Brand near Freiberg (G.D.R.)
Size: 12 cm × 7 cm × 4 cm

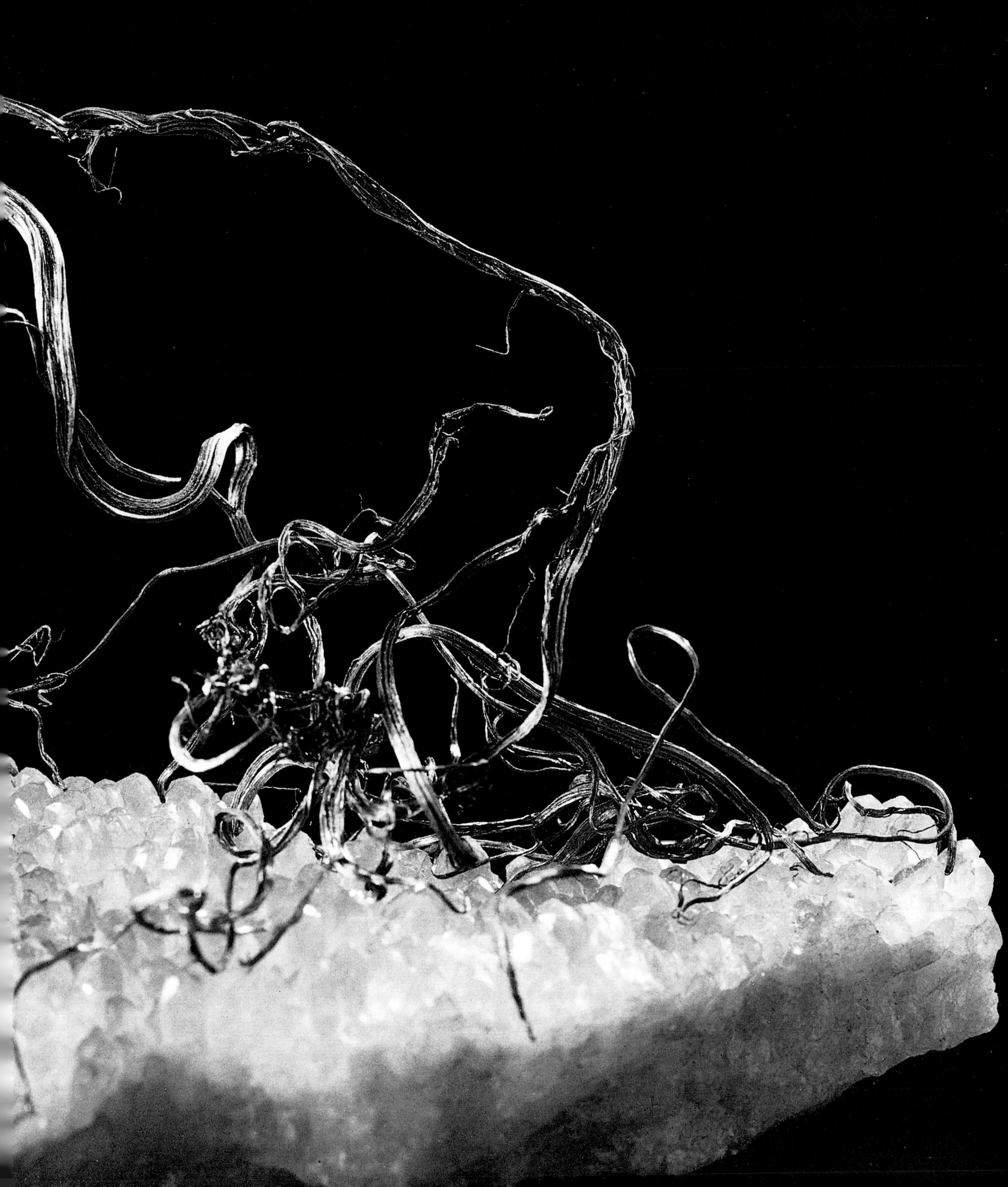

Pyrargyrite

The unimpressive-looking, rare silver ores have long been the subject of scientific investigation. They were classified and named according to their external features and their composition. A G WERNER discovered that the difference between dark and light ruby silver ores depended on the content of antimony or arsenic. F A BREITHAUPT then established the distinction between the two varieties, classifying them as antimony silver ore and arsenic silver ore respectively. In 1831 R GLOCKER named the dark ruby silver pyrargyrite, while the light ruby silver was named proustite, after the French chemist, J L PROUST.

The prismatic crystals with a metallic lustre are mostly opaque and black. Only the thin parts occasionally have dendritic forms with delicate branchings, and are translucent red.

Magnificent specimens come from the mines near St Andreasberg in the Harz (Federal Republic of Germany) and from Freiberg (German Democratic Republic). Localities with occurrences of equal value are at Zacatecas and Guanajuato in Mexico, and it occurs in many of the silver mining districts of the Rocky Mountain States of the United States of America. It is also found in the silver district of Cobalt, Ontario, Canada.

Silver, Freiberg (G.D.R.)
Size: 9cm × 8cm × 6.5cm

Silver (cut surface with inscription)
Freiberg (G.D.R.)
Size: 9cm × 8cm × 6.5cm

Silver, Marienberg, Erzgebirge (G.D.R.)
Size: 6cm × 5cm × 3cm

Pages 70/71
Argentite
Gersdorf near Rosswein (G.D.R.),
found 1843
Size: 6cm × 19cm × 3cm

Acanthite, Brand near Freiberg (G.D.R.)
Size: 0.5cm × 5cm × 0.3cm

Pyrargyrite with pyrite and ankerite
Zug near Freiberg (G.D.R.)
Size: 4.5cm × 5cm × 4cm

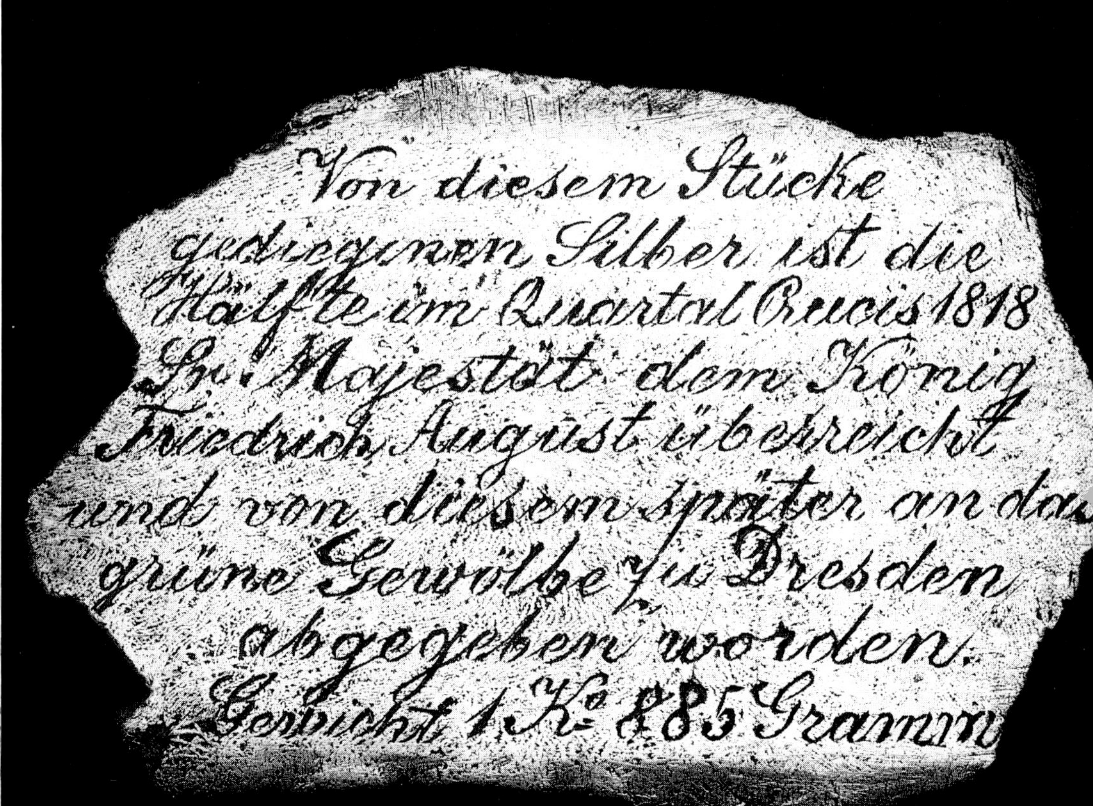

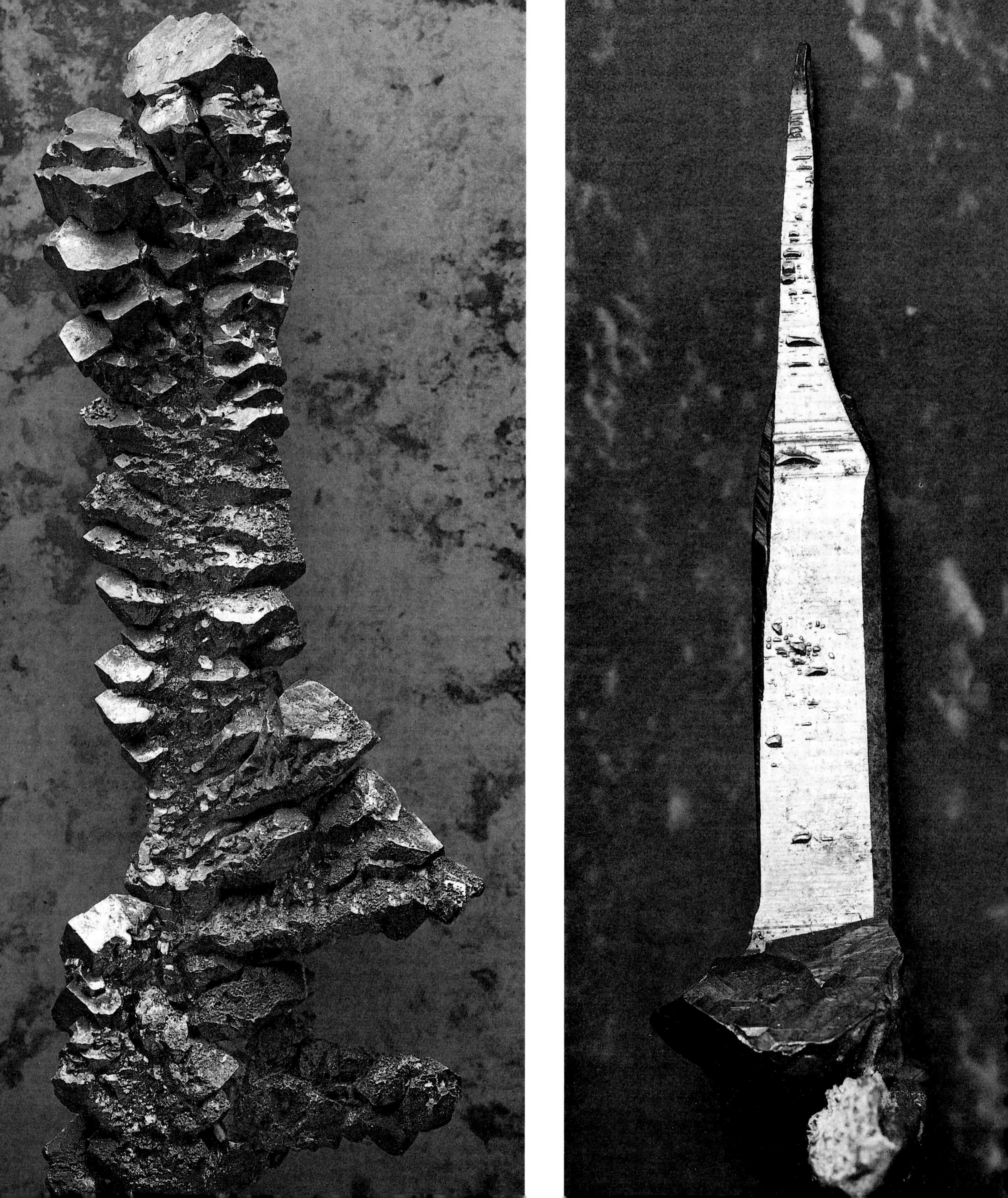

Argyrodite

In 1820 the first samples of argyrodite were
found in the Simon Bogner mine near Freiberg, and were described by F A BREIT-
HAUPT as a new type of silver which he called
Plusinglanz. Sixty-five years later, A WEIS-
BACH analyzed a rich silver ore (almost 76 %
silver) from the Himmelsfürst site at St Mi-
chaelis near Freiberg, and named it argyrodite. This mineral, with a bright metallic
lustre, forms botryoidal or warty masses, of
black to steely-grey colour, with a bluish
violet shimmer in reflected light. The presence of marcasite tends to cause the specimens to decompose, giving an iron sulphate
efflorescence.

On 6 February 1886, C WINKLER, a chemist at the Freiberg Mining Academy, discovered the element, germanium. At the
time he wrote: "After several weeks of exhaustive investigation, I am today able to
state that argyrodite contains a new element
very similar to antimony, but quite distinct
from it, which is to be named 'Germanium'."

The properties of the new element had
already been predicted by D I MENDELEYEV
in 1872, on the basis of the classification of
elements in the periodic table. Since germanium follows silicon in Group IV, it was
known, until the time of its discovery as
'ekasilicon.'

Apart from being found in Freiberg,
German Democratic Republic, argyrodite
has also been found in Potosí and Aullagas
in Bolivia. Both occurrences have long since
been exhausted. As there is no way of replen-
ishing it, quite apart from its historical interest, the value of argyrodite is rising all the
time.

Argyrodite
Brand near Freiberg (G.D.R.)
Size: 11 cm × 7.5 cm × 3 cm

Bournonite with calcite on quartz
Bräunsdorf near Freiberg (G.D.R.),
found 1846
Size: 1 cm × 2 cm × 0.8 cm

Miargyrite

Only a few private collections contain this rare silver mineral. The richest finds date back to the middle of the last century. Before that, miargyrite was not recognized, and was mistaken for ruby silver ore because of its association with other antimony (berthierite and kermesite) and silver minerals, especially pyrargyrite. The name miargyrite was chosen when the analysis revealed how little silver it contained by comparison with ruby silver ore. The steel-grey crystals from the Neue Hoffnung Gottes mine at Bräunsdorf near Freiberg have a near adamantine, metallic lustre and occur predominantly in cavities with quartz. The thick tabular, richly facetted crystals frequently have deeply striated faces. Miargyrite is also found in the famous St Andreasberg silver deposits (Federal Republic of Germany), in Příbram in Czechoslovakia, and at Zacatecas in Mexico.

Bournonite, Wheel Ore, Endellionite

Well crystallized bournonite occurs in only a few deposits, and consequently they are highly-prized rarities in collections. The regular intergrowth of several, individual, chisel-like crystals of bournonite produce a characteristic cog-wheel shape. Such crystal forms occur frequently in the mines near Cavnic in Rumania and the miners of Transylvania named them 'wheel ore.'

The first description of bournonite was given by the French mineralogist, Count JL de Bournon in 1813. He described the mineral from St Endellion in Cornwall as a sulphide of lead, antimony and copper, and named it endellione. Later, R Jameson proposed the name bournonite in honour of the Frenchman.

Quartz, var. rock crystal, cut and polished, showing inclusions of rutile needles
Minas Gerais (Brazil)
Size: 8cm × 6cm × 3cm

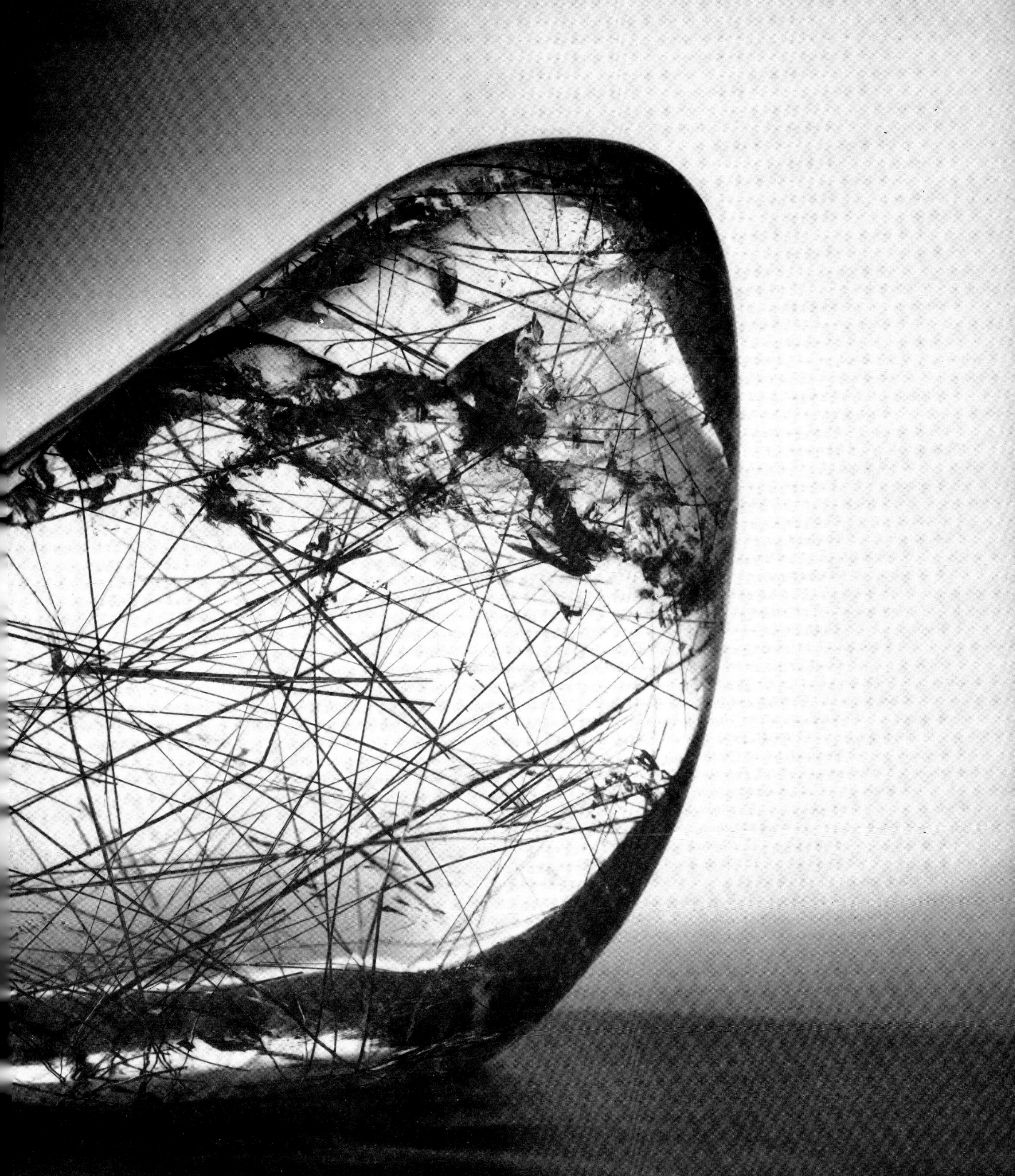

The finest crystals were found in Neudorf and Wolfsberg (in the Harz), near Horhausen in Westphalia (Federal Republic of Germany), in Příbram in Bohemia (Czechoslovakia), Cavnic and Săcăramb in Rumania, as well as at Wheal Boys, St Endellion and Liskeard in Cornwall. In Bolivia it occurs at the Pulacayo mine near Huanchaca, Potosí, and in the United States of America is found in Arizona, Montana, Utah and Nevada.

Quartz

Quartz is one of the commonest minerals found in the crust of the Earth. As early as the fourteenth century it is recorded that this hard, compact mineral made the Harz miners' work more difficult, and since it was barren it yielded no profit. The minerals associated with ores are known as 'gangue,' and quartz is probably the commonest gangue mineral. Chemically, quartz is composed of silicon and oxygen, and may contain many impurities producing numerous varieties, differing in colour. All varieties of quartz have certain features in common. The fracture is conchoidal, demonstrating the absence of cleavage. Hardness is relatively high, and all minerals that scratch quartz are regarded as gemstones in the strictest sense. Another characteristic is its durability, due mainly to its resistance to the common acids; though hydrofluoric acid and alkalis (caustic potash and caustic soda) cause perceptible corrosion in quartz. The characteristic form is six-sided crystals, the prism faces of which frequently have striations running across their length.

The many varieties of quartz can be classified according to colour and structural characteristics.

Quartz, var. smoky quartz, twisted crystal (Gwindel—half-open type)
Maderan valley near Amsteg, Uri (Switzerland)
Size: 10cm × 14cm × 4cm

Quartz, var. rock crystal, with inclusions of actinolite needles and hematite
St Gotthard region (Switzerland)
Size: 4cm × 9cm × 4.5cm

Rock crystal is quartz in its purest, colourless form. Single crystals more than one metre in size and weighing several hundredweight have been found in Brazil at Minas Gerais, Goiás and Rio Grande do Sul. Other localities with a highly profitable production of quartz are the island of Madagascar and Arkansas, United States of America, which have also produced huge crystals and magnificent specimens for collections. In Europe, rock crystal occurs widely in the Alps, in Switzerland, Austria, Italy and France where it is found mainly in joint cavities. Fine groups of clear crystals come from Frizington, Cleator Moor, and from Alston Moor in Cumbria, and examples of these can be seen in many museums.

The purity of rock crystal is often compared with diamond, although the hardness and refractive indices of quartz are considerably less than that of diamond. Nevertheless, the small, colourless, water-clear, wellnigh perfect quartz crystals are set on a par with diamond, which is used in the popular names of some of them. Examples are the Herkimer diamonds from Herkimer County, New York, United States of America, and the Marmaros diamonds from Rumania, but the rock crystal from the Carrara marble in Italy is also of great purity. Quartz often contains inclusions and may be milky. During growth, water, solutions and gases may be included, which may make the crystal completely or partially cloudy. Cloudy quartz crystals, which are quite common, are called milky quartz. These are of interest to the collector only if the crystals are well formed.

Quartz with coating of gilbertite
Ehrenfriedersdorf, Erzgebirge (G.D.R.)
Size: 10cm × 23cm × 8cm

Quartz with traces of pyrite
Annaberg, Erzgebirge (G.D.R.)
Size: 19cm × 15cm × 9cm

Pages 80/81
Quartz, var. rock crystal
Bourg d'Oisans, Dauphiné (France)
Size: 15cm × 17cm × 13cm

Quartz, var. rock crystal with zoned
structure and inclusions of clay between
the layers. Poretta near Bologna (Italy)
Size: 6cm × 5cm × 4.5cm

79

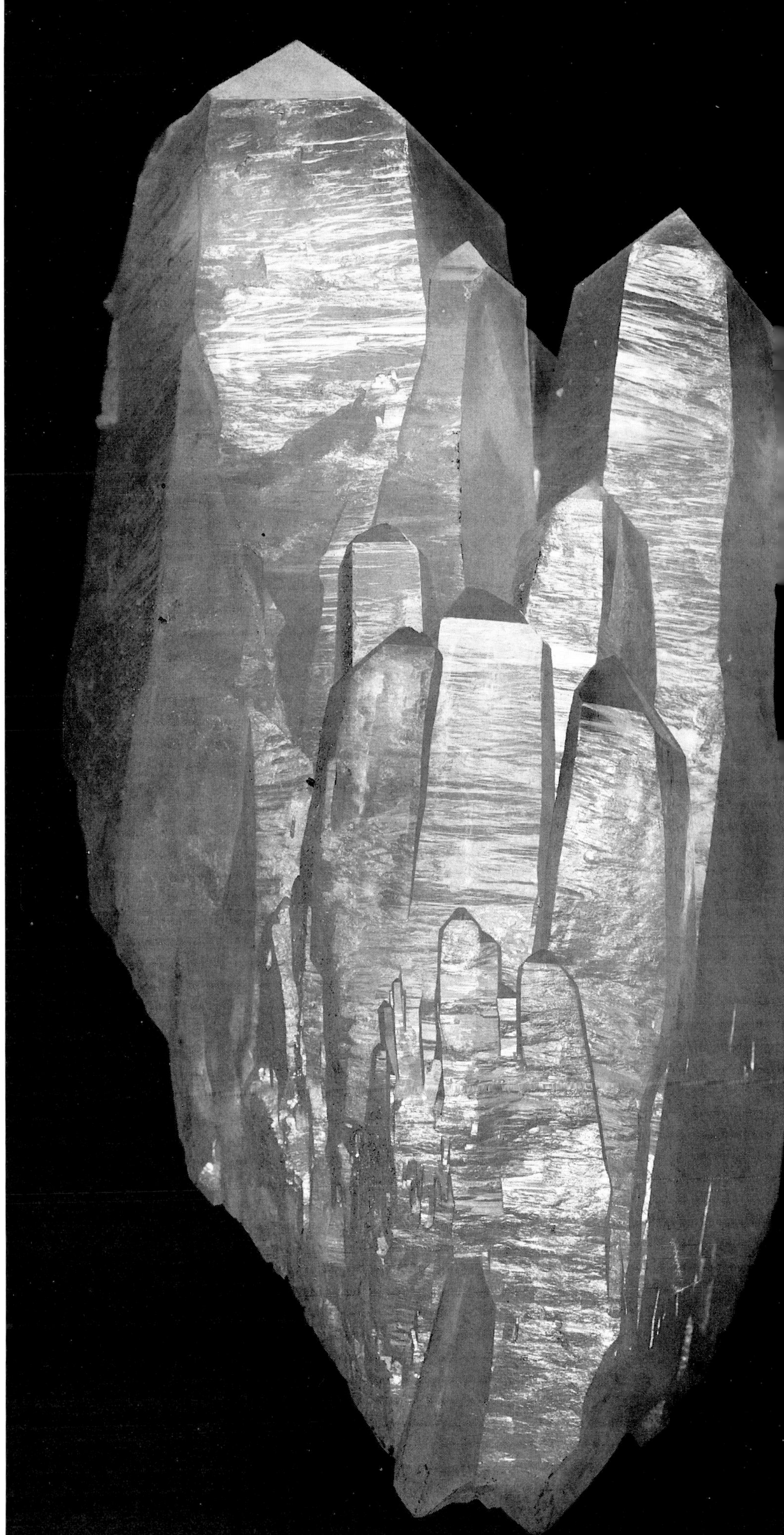

Considerably more attractive forms are those in which other minerals have been included. The straw-yellow sheaves of rutile needles or the bronze-coloured clusters of rutile needles included in rock crystal are among the special attractions of a private collection, as are inclusions of green and black tourmaline. Alpine rock crystals may have inclusions of dark green, spear-shaped epidote, and leek-green, foliated or scaly chlorite. Much rarer inclusions are cassiterite, hematite, calcite, dolomite and siderite.

A rule of thumb used by commercial mineral collectors is that smoky quartz is found in the Alps at an altitude above 2,500 metres and only in certain rocks (granite). Light to dark brown quartzes, when heated to a temperature of 260 °C will alter to clear rock crystal. The process can be reversed by subjecting crystals to radio-active radiation.

Smoky quartz crystals from Switzerland occasionally produce 'Gwindel' formation. These are twisted quartz crystals, which develop by parallel growth along a common axis of identical, individual crystals. The slight displacement of the individual crystals causes the faces of the resulting crystal to appear twisted. Smoky quartz is found in the Cairngorm Mountains in Scotland and is widely used in jewellery worn with Scottish national dress. Fine smoky quartz crystals are also found in the Mourne Mountains, Ireland.

A find made in the Tiefengletscher in Switzerland in 1868 was outstanding. During a period of eight days, a mineral pocket at an altitude of 3,100 metres yielded 200 ctr of mostly black quartz crystal (morion). The largest were each given local names and, each weighed more than 100 kilos.

Other coloured varieties of quartz, such as amethyst, citrine and rose-quartz, are used mainly in jewellery. Structural variations frequently occur in quartz. These are structural defects that develop as the crystal grows. They take the form of distortions of the crystal structure, and are influenced by the prevailing crystallization conditions such as temperature and pressure.

Quartz, var. agate
Uruguay
Size: 8cm × 6cm × 4cm

Quartz, silica (horn stone), pseudomor-
phous after calcite (found by Breithaupt
in 1846)
Schneeberg, Erzgebirge (G.D.R.)
Detail

Rock crystal with rutile and siderite in-
clusions, Madagascar
Diameter: 4 cm

In skeletal quartz, the edges of the crystals grow more quickly than the centres of the faces, so that faces are depressed. A particular type of skeletal growth is known in German as 'window quartz' and examples occur at Poretta in the province of Bologna in Italy.

Zoned structure is produced by interruptions in the growth of the crystal. Inclusions between the layers emphasize this zonation, and the individual zones are often easily seen in clear crystals. The rhombohedral faces which terminate quartz crystals are particularly prone to such zonation and often produce 'capped' quartz and are closely related in origin to 'sceptre' quartzes.

85

Chalcedony is composed of minute quartz crystals, which cannot normally be resolved but under the microscope. Chalcedony absorbs water which is held between the minute quartz crystals. It forms mainly botryoidal, stalactitic or mamillated masses which can be cloudy or translucent, and occur in a wide range of colours, with grey, brown and reddish tones predominating.

Agate is chalcedony that is characterized by a fine banding in different colours and the finest agates come from Uruguay and Brazil, though small ones can be found at a number of localities in Scotland.

Jasper is deep red or brown, and opaque and has, like chalcedony, a conchoidal fracture.

Fluorite (water-clear, octahedral cleavage fragment)
Oltscherenalp, Brienz, Berne (Switzerland)
Size: 3.5 cm × 5 cm × 3.5 cm

Smoky quartz with zinnwaldite
Zinnwald, Erzgebirge (G.D.R.)
Size: 12 cm × 20 cm × 16 cm

Fluorite shows an impressive variety of colours. Colourless crystals are rare; strong colours predominate. Inky-blue fluorite is a particular favourite, but the honey-yellow, grass-green, deep purple and pink specimens are much admired. Crystals are mostly translucent, and occasionally transparent, which enhances the different colours shown occasionally by single crystals from some localities; the centre generally being paler (yellow or pink) than the outer parts of the crystal (blue or green). Fluorite shows most spectacular colouring in ultra-violet light. The strength of the fluorescent colour is almost without equal. One distinct disadvantage of this mineral, however, is that the natural colours fade. The effect of direct sunlight, or storage under conditions of low humidity will reduce the originally brilliant colours to a milky, greyish-white.

The commonest form is the cube but the octahedron also occurs. The faces of the cube usually show the familiar 'moist' vitreous lustre, while the faces of the octahedron often appear matt. It has a perfect octahedral cleavage.

The name fluorspar is closely linked to its use and to one of its essential properties. It is used in the smelting industry as a flux to liquefy slag, and to separate sulphur and phosphorus from metallic melts.

Fluorite is also used in the production of enamel and achromatic lenses, for etching glass, and occasionally in jewellery and for decorative purposes.

There are many localities in England which have produced fine fluorites. Perhaps the most widely known is 'Blue John' from Derbyshire which has been worked since Roman times and from which objects such as vases and goblets were carved. It is also found at Cleator and Alston Moors in Cumbria, and Weardale in Durham, and it is abundant at many of the mining localities of Cornwall. It is a common gangue mineral in many of the lead veins of the north of England.

Fine fluorite is found at La Gollada near Oviedo in Spain (purple-blue), at Rosiclaire and Salem of Illinois and Kentucky respectively in the United States of America, as well as at Muzquiz and Coahuila in Mexico and many other localities.

Calcite

The chemical formula for pure calcite is $CaCO_3$. Depending upon the temperature and pressure, calcium carbonate can crystallize in the form of calcite or aragonite. Apart from calcium, other elements such as iron, magnesium, strontium and lead may be present and may influence the colour of the mineral, while their presence can also produce 'mixed' crystals having particular names such as mangano-calcite or plumbo-calcite. Besides the elements listed above, further substances can be incorporated into the calcite as impurities during the crystallization.

Besides the wide range of colours that extends from white, through red, yellow and brown to black, the variety of crystal forms shown by calcite is quite remarkable. There are several hundred different forms of the most richly-facetted and well-developed crystals, calcite being the mineral with the greatest variety of form. So far seventy-five different rhombohedra and over two hundred scalenohedra have been observed.

A useful aid in recognizing a distorted crystal is to look for the fine, distinct cleavage on faces that are parallel to the main rhombohedron. The variety of crystal habit can be simplified considerably by dividing them into main groups of rhombohedral, scalenohedral, prismatic and tabular crystals. The variety of crystal habit is increased even further by twinning, and several twin laws are known. The heart-shaped twins from Egremont in Cumbria are particularly striking.

Calcite is very soluble in carbonic acid, and can be reprecipitated from this solution. The conical and columnar shapes known as stalactites and stalagmites are formed by the precipitation of calcite from dripping water in caves. The compact, fibrous structure of such calcite is often visible on broken surfaces.

Fluorite, translucent, blue-yellow
Halsbrücke near Freiberg (G.D.R.)
Size: 17cm × 10cm × 7cm

Fluorite with pyrite
Halsbrücke near Freiberg (G.D.R.)
Size: 10cm × 10cm × 6cm

An optical phenomenon that can be clearly
demonstrated using water-clear (optically
pure) calcite, earned it the name 'double
spar.' Handwriting or a mark of some sort
drawn on paper will appear as a double image,
when viewed through calcite. This effect was
discovered in 1669 by E.BARTHOLINUS, with
calcite from Iceland. It is caused by the division of a beam of light into two rays, that
pass through the calcite at differing speeds.
The difference in the speed of the two rays,
known as double refraction, is, in calcite, so
great that it is visible to the naked eye. 'Iceland spar' has been found also in Namibia
and in the Soviet Union.

Calcite is widely distributed and occurs
also massive. Some mountain ranges, like
the Calcareous Alps are formed substantially
of calcium carbonate in the form of the rock
limestone. Particularly fine specimens come
from Alston Moor and Egremont in Cumbria,
from Matlock in Derbyshire, Weardale in
Durham, near Furness in Lancashire, and a
number of localities in Cornwall. Stalactites
and stalagmites are found in many caves in
limestone areas, particularly the Pennines.

Many deposits of massive calcite are worked
commercially, the construction and smelting
industries being the main consumers. Large
quantities of limestone including chalk,
are required in the production of cement
and lime (calcium oxide). It is used in blast
furnaces as a calcareous flux in the smelting of ores. In agriculture and gardening,
lime is used to neutralize acid soils.

Calcite
Freiberg (G.D.R.)
Size: 10cm × 6cm × 4cm

Siderite (Chalybite, Iron Carbonate)

'White ironstone' can be distinguished from other iron ores by its pale, yellowish-brown colour. When exposed to air and in water, siderite rapidly alters to limonite, the rust-brown colour being a characteristic sign of the presence of trivalent (ferric) iron. This is why siderite always forms in the absence of oxygen. The crystals are rhombohedral and have perfect rhombohedral cleavage. Scaly, curved lenticular or saddle-shaped crystal faces are produced as a result of the granular structure of siderite, and are characteristic of certain deposits. Because of the close connection between siderite, rhodochrosite, and calcite, these minerals form solid solution series. Manganese-bearing siderite has a bluish-black colour on weathering. Spherical, radiating forms are called sphaero-siderite,

Calcite, heart-shaped twin with hematite bloom
Egremont, Cumbria (Great Britain)
Size: 9 cm × 8 cm × 5 cm

Calcite with pyrite and baryte
Gersdorf near Rosswein (G.D.R.)
Size: 17 cm × 13 cm × 4 cm

and those with inclusions of clay, quartz and carbon are called clay ironstone.

Of the many known localities, the following are just a few that have yielded especially beautiful crystals. Particularly fine specimens of siderite have been found at Tavistock in Devon, and were found in many of the old Cornish mines, notably those at Camborne, Bodmin, Redruth and St Austell. Sideritic iron ores were formally of considerable economic importance in England and Scotland. There are further important deposits at Cavnic in Rumania, Příbram in Bohemia (Czechoslovakia), Travetsch in Switzerland, Allevard, Isère in France, and near Ouro Prèto, Minas Gerais in Brazil. The largest deposit is at Erzberg near Eisenerz in Austria, where it has been worked as a source of iron since Roman times.

Calcite with sulphur
Agrigento, Sicily (Italy)
Size: 13cm × 17cm × 10cm

Calcite with pyrite on quartz (well-formed
scalenohedral crystal)
Niederrabenstein, near Karl-Marx-Stadt
(G.D.R.)
Size: 6cm × 6cm × 4cm

Pages 96/97
Calcite, crystallized sandstone
Badlands, North Dakota (U.S.A.)
Size: 10cm × 9cm × 4cm

Calcite, calcareous concretion
Sunderland, Durham (Great Britain)
Size: 10cm × 11cm

Pages 98/99
Calcite
Kamsdorf near Saalfeld, Thuringia (G.D.R.)
Detail

94

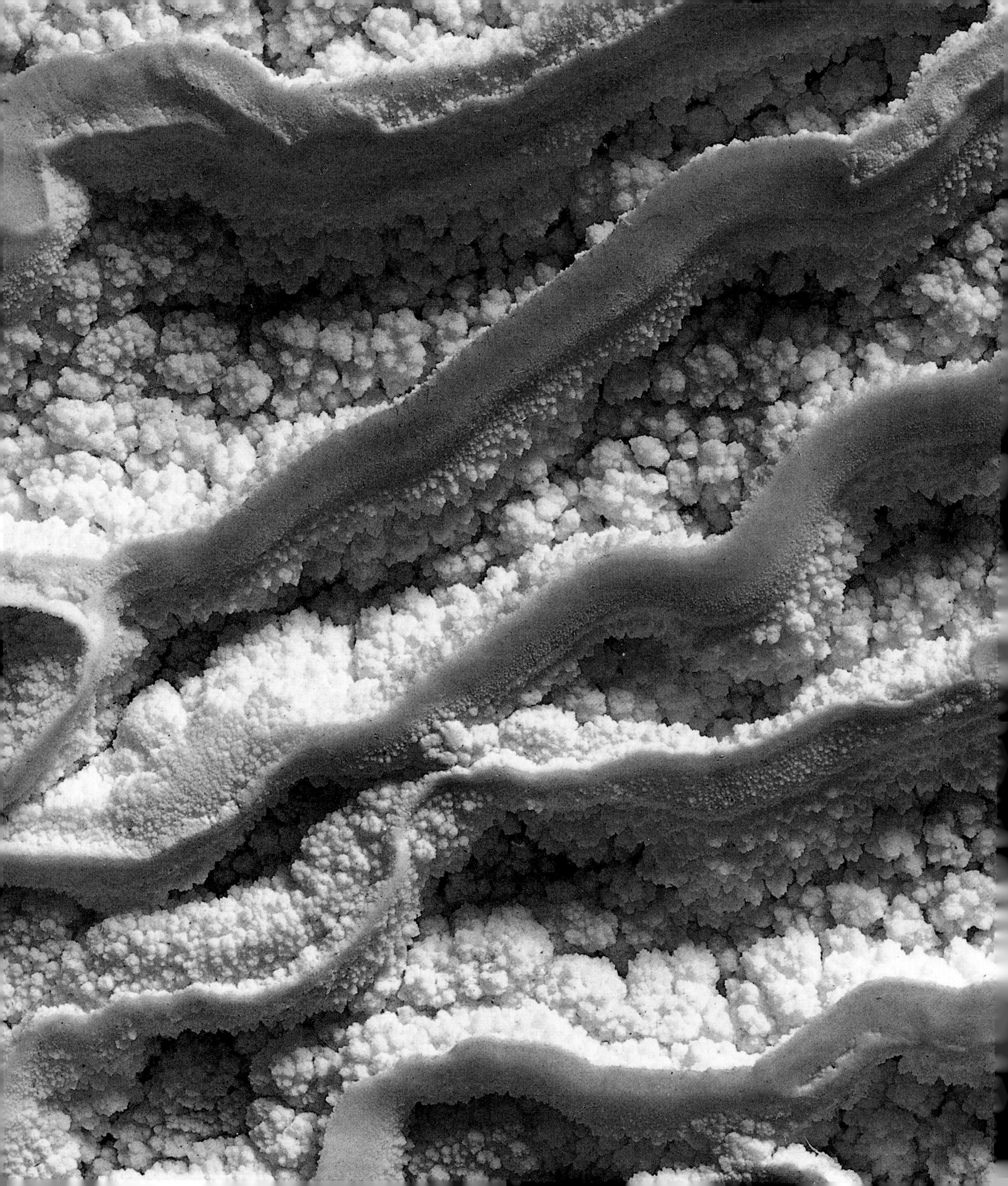

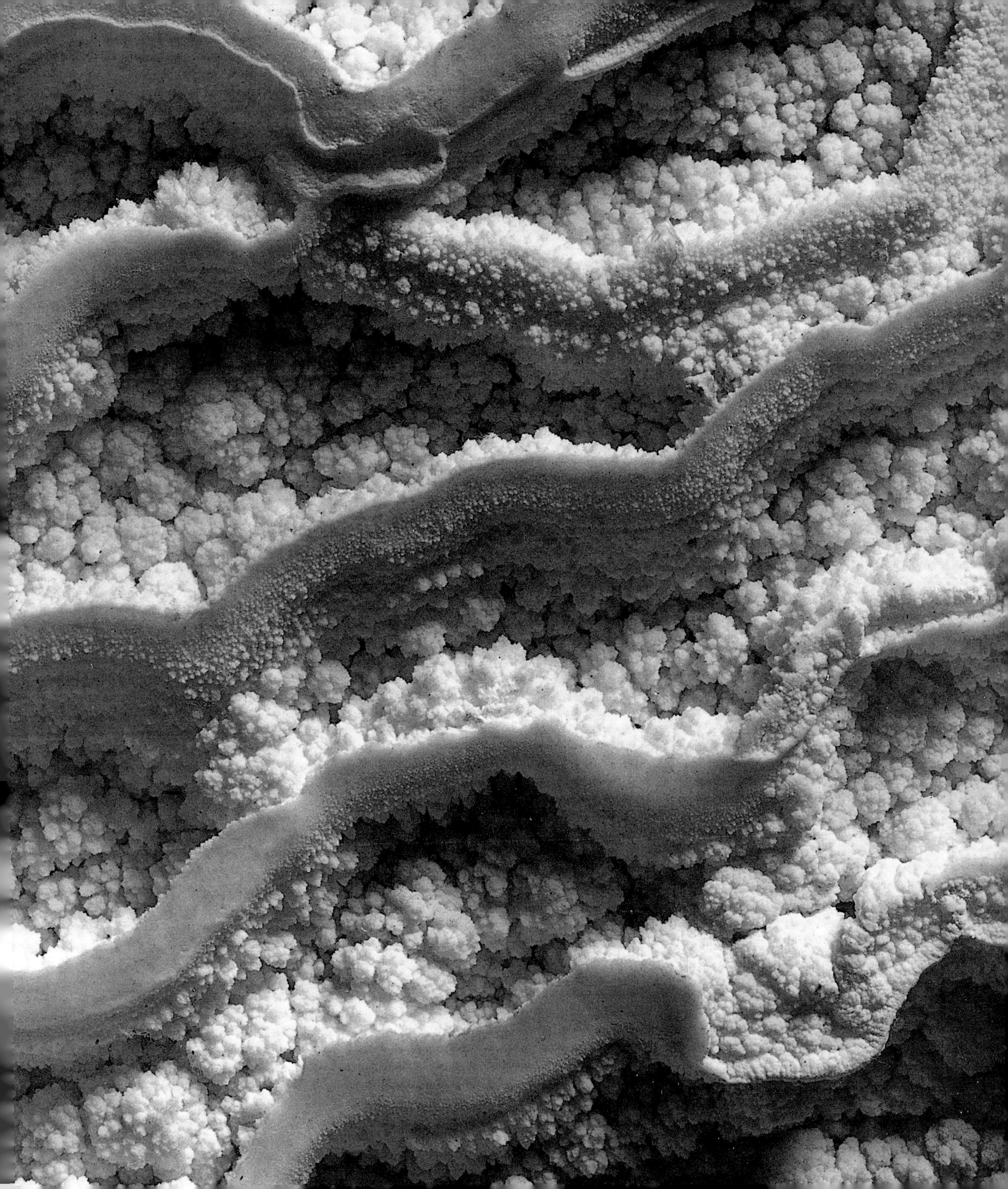

Rhodochrosite

The colour of a mineral can be so fascinating, that this alone accounts for its popularity and helps in its recognition. The soft pink tone of rhodochrosite is produced by bivalent manganese. The name derives from the Greek words, *rhodon* meaning rose, and *chrosma*, meaning colour. The gradations in colour, the change from pale to dark pink, as well as the variety in the banding and figuring produce interesting patterns, and give rise to speculation as to the origin of the mineral. Like malachite, the forms are strongly reminiscent of agate. Rhodochrosite is a carbonate, usually forming bulbous masses; it occurs in hydrothermal veins and in metamorphosed sedimentary rocks. Crystals are rare, but are rhombohedral in form (trigonal system), and have a perfect rhombohedral cleavage. Though easy to work, its inferior hardness restricts its use as a gemstone, as polished rhodochrosite does not stand up well to heavy wear. It is ideal for making into pendants, platters or bowls.

The best quality of rhodochrosite comes from deposits in Capillitas, Catamarca and San Luis in Argentina. The most superb crystallized specimens of ruby-red, transparent rhodochrosite on quartz with pyrite and tetrahedrite were found in 1965 when a tunnel was being driven in Alma, Park County, Colorado, in the United States of America. Other well known localities are Cavnic in Rumania, as well as Silverton, Cripple Creek and Butte in the United States of America. Some rhodochrosite also came from St Just in Cornwall.

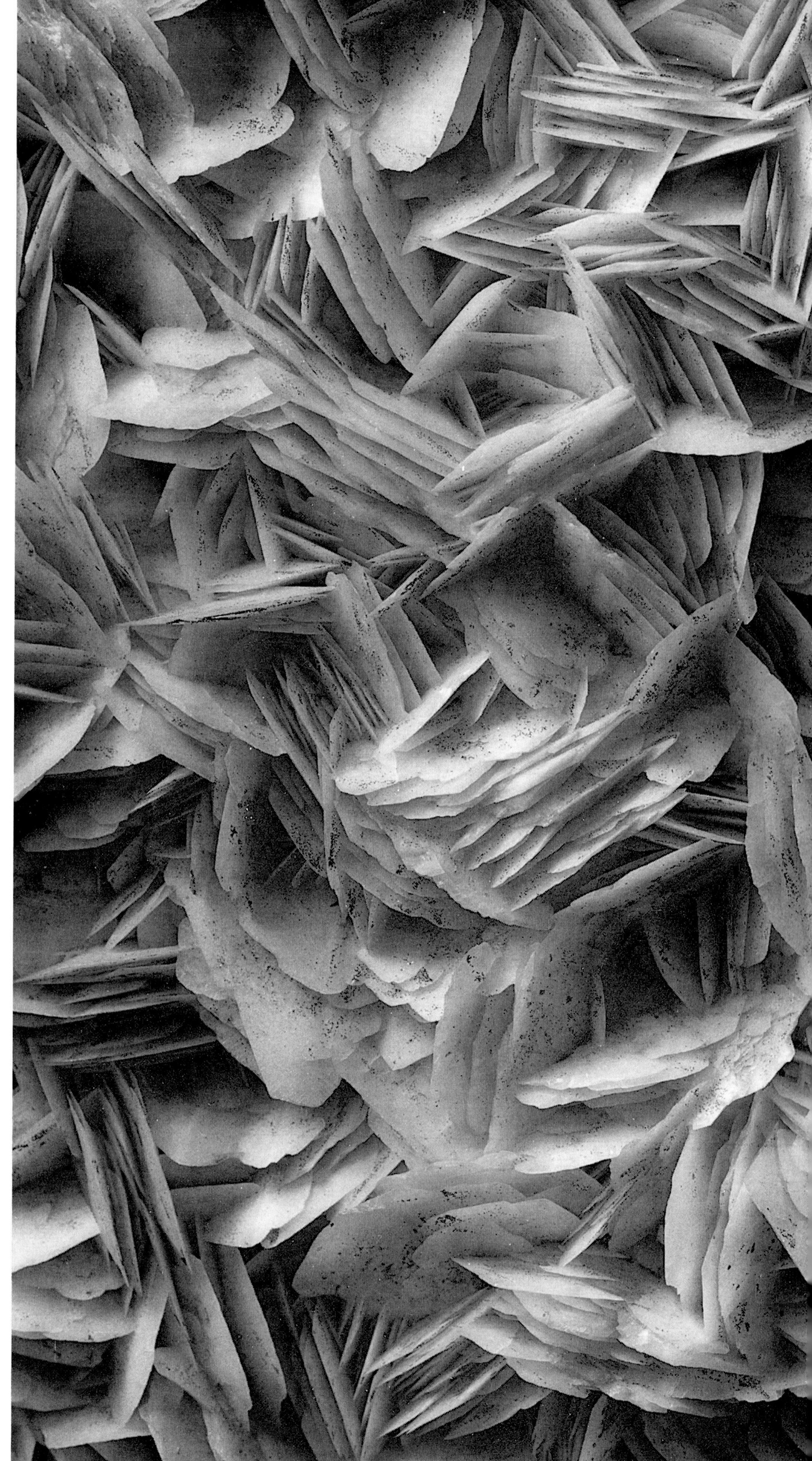

Calcite, thin platy crystals
Schneeberg, Erzgebirge (G.D.R.)
Detail

Pages 102/103
Ankerite, pseudomorph after calcite
Freiberg (G.D.R.)
Detail

Rhodochrosite with quartz
Cavnic near Baia Mare, Transylvania
(Rumania)
Size: 15cm × 12cm × 3.5cm

Baryte (Barytes, Heavy Spar)

Variety of crystal shape and habit, and wealth of colour are often the reason for specialist collections of certain minerals. Thus private collections of baryte have been made, in which the collection of specimens was not restricted to one locality. Fluorite, calcite and quartz all show wide variation in colour, form and habit, and prompt collectors to specialize.

These minerals deservedly take pride of place in the exhibits displayed by mineralogical museums. In order to demonstrate their most important and striking forms, it is important to be able to draw on a very large number of specimens, which are made the more attractive by the variation in colour they display.

Over two hundred different forms of baryte have so far been recorded. Tabular crystals predominate, often forming a fan-like arrangement as a result of parallel growth. The colour ranges from yellow, brown, red, green and blue to black. Colourless crystals are comparatively rare. Rosette-like concretions are found in the Tertiary sands of the Wetterau in Hessen (Federal Republic of Germany). Baryte roses have been found at Rockenberg (Federal Republic of Germany), weighing up to 5 kg.

Scientific interest in baryte was awakened in 1630, with the observation of V Cascariolo, a shoemaker from Bologna, that when baryte from Monte Paterno, near his home town, was heated together with organic matter, it continued to glow in the dark afterwards. This property of baryte, known as phosphorescence, is still used today in the manufacture of illuminated watch dials. Ground baryte is also used in the muds that are pumped into oil wells to prevent blow-outs and to carry to the surface the rock drillings. It is used in this way because of its very high specific gravity.

Baryte
Schneeberg, Erzgebirge (G.D.R.)
Detail

Rhodochrosite with pyrite, marcasite and sphalerite
Capillitas, Catamarca (Argentina)
Detail

The name, heavy spar, derives from the high specific gravity of baryte and it was for a long time regarded as a heavy form of gypsum. The new 'heavy earth' was detected in heavy spar by two Swedish chemists, K W Scheele and J G Gahn, at almost the same time in 1774, but independently of each other. The French named the new earth, baryte (in Greek *baryte* means 'heavy'). It was not until 1901 that the element barium was extracted from baryte.

Of the many known deposits of baryte, apart from those named above, particularly fine crystals have been found in Cornwall at a number of localities, also from several localities in Cumbria, Shropshire and Derbyshire; very large crystals, including one of about 50 kg, came from the Dufton lead mine in Cumbria.

Baryte with chalcopyrite
Dreislar near Medebach, North Rhine-
Westphalia (Federal Republic of Germany)
Detail

Crocoite

This lead mineral owes its name to its striking red to orange-yellow colour, *krokos* being Greek for saffron. In 1797 L N VAUQUELIN detected the element, chromium, in this ore, which, being present as the chromate radical, produces the vivid colour of crocoite. The first material to be investigated came from the Beryesovsk mines in the Urals, and was described by M V LOMONOSOV in 1766. The long prismatic crystals have many faces and are often hollow. Acicular crocoite often occurs in bundles in cavities, or lies flat against the cavity walls.

By far the most beautiful specimens of hyacinth-red, transparent crocoite were found in Tasmania in 1895.

Other well known crocoite localities are in Brazil (Goiabeira), southern Rhodesia (Penchalonga, Umtali district), the Island of Luzon in the Philippines, and the Cavnic region in Rumania. In addition a discovery has recently been made in the neighbourhood of Glauchau in the German Democratic Republic. Here, prismatic crystals, up to 3 cm in length, mostly arranged in bundles, occur in cavities in serpentinite.

Baryte
Wiesloch, Baden-Württemberg (Federal Republic of Germany)
Size: 15 cm × 6 cm × 11 cm

Baryte with dolomite and hematite
Frizington, Cumbria (Great Britain)
Size: 9 cm × 8 cm × 6 cm

Wulfenite
Mežica, Karawanken, Slovenia (Jugoslavia)
Size: 10cm × 7cm × 6cm

Wulfenite

The bright lemon-yellow colour of this lead
mineral results from the presence of molybdenum, and A G WERNER had in mind both
the yellow colour and main metallic constituent, when he named it 'yellow lead ore'
(German, *Gelbbleierz*). Apart from the mo-
lybdate wulfenite, the striking bright red
colour is shown also by chromates (crocoite)
and vanadates (vanadinite). It is interesting,
in this respect, that these elements (Mo, Cr,
V) occur close together in the periodic table
(Groups V and VI), which expresses the
similarity of their properties. An Austrian
baron, ABBÉ VON WULFEN, of Klagenfurt was
the first to describe the mineral, referring to
it as 'Carinthian lead spar,' and in 1845,
W HAIDINGER suggested it should be named
wulfenite in his honour.

Specimens from Bleiberg in Carinthia
(Austria) showing groups of thin, tabular
crystals, are particularly attractive, as is the
wulfenite coloured red by chromate from
Yuma County, Arizona, in the United States
of America.

There are further well known deposits of
wulfenite in the lead ore deposits in Jugoslavia: orange to honey-coloured pyramidal
crystals have been found at Mežica and
Crna. In Příbram, Bohemia (Czechoslovakia),
smoky-grey, tabular wulfenite has been
found in cavities in galena.

Mimetite on porous limonite
Santa Eulalia, Chihuahua (Mexico)
Detail

Wulfenite with hematite on dolomite
Los Lamentos, Chihuahua (Mexico)
Detail

Mimetite

It has already been mentioned above that many lead ores are brightly coloured, and this led A G WERNER to classify them according to colour. He distinguished white, red, yellow, brown, green, blue and black lead ores. The crystal forms of mimetite are deceptively similar to those of pyromorphite, which led F A BREITHAUPT to choose the name, mimetite, *mimetes* being Greek for 'imitator.'

Excellent specimens have been found in the Johanngeorgenstadt deposits in the German Democratic Republic. The prismatic crystals occur in groups, resembling rosettes or buds. Doubly-terminated crystals or 'sceptre' growths are not uncommon, and can be over 1cm long. Barrel shapes with curved faces have also been found which, like the other varieties, have grown on ferruginous quartz. Their colour ranges from honey-yellow, greenish-white to brown.

112

The finest specimens of mimetite were found in Namibia—occurring as large (over 3 cm), yellowish brown, opaque crystals as well as transparent, bright yellow ones, of perfect quality.

Mimetite occurs as a secondary mineral in lead ores, especially those containing arsenic. Another, highly-prized variety is orange-coloured campylite, found in Cornwall and Cumberland. In this mineral, arsenic has largely been replaced by phosphorus.

The prize specimens most recently acquired by collections are sulphur-yellow, cauli-flower-like mimetite from Santa Eulalia in Mexico, and exquisite, amber-coloured crystal specimens from mines in the neighbourhood of Nerchinsk, eastern Transbaikal in the Soviet Union.

Mottramite

This is a very rare mineral and has been found in the form of beautiful crystals in Namibia. Its lustre is particularly striking, and the pointed pyramidal formations resemble clusters of growing water plants. Botryoidal and crusty aggregates are more common. The colour derives from its copper content, with brown to black tones predominating, while olive green shades occasionally occur. Major deposits in Abenab (Namibia) and Kabwe (formerly Broken Hill, Zambia) are worked as a source of vanadium.

Legrandite

Only a few collections possess well-crystallized specimens of legrandite. This rare mineral was discovered in 1920 by LEGRAND near Lampazas, Nuevo León in Mexico. Several superb specimens appeared on the mineral market at the end of the 'sixties', but demand soon exhausted the supply.

Crocoite with limonite
Dundas, Tasmania (Australia)
Size: 12 cm × 9 cm × 3.5 cm

The orange-yellow prismatic crystals on these exclusive specimens are 3 cms long, and occur mostly in fan-shaped clusters. The pale colour of the legrandite is in striking contrast with the deeply-fissured matrix of dark brown limonite. Legrandite also occurs in association with sphalerite, pyrite and siderite.

Adamite

The yellowish green of adamite forms a striking contrast with the rich brown of the limonite matrix. The rosette-like hemispheres of light, shining adamite rest on cellular limonite, and the radiating, fan-like structure of the adamite produces an undulating and fissured surface. The lamellar formation is plainly visible on broken surfaces of the hemispherical rosettes. The depth of colour varies, with the greenish tone of the core fading to a paler colour at the periphery. The individual crystals are translucent. All these features make adamite a highly-prized mineral in collecting circles.

Adamite was first found together with cerargyrite and quartz in Chañarcillo in Chile. The samples submitted for investigation by G J ADAM of Paris, led to the discovery of the new mineral, which was named after him.

In the past the demand for adamite from Mapimi Durango in Mexico was enormous.

Copper and cobalt may partially replace the zinc in adamite, and give the mineral a deeper green or pink tone. Both sub-varieties have their own species name (cuproadamite and cobaltadamite) and are typical of the deposits at Cap Garonne in the Var département in France.

For a long time the beautifully crystallized specimens of yellow, green or blue adamite from the zinc deposits in Laurion in Greece have served as the standard for lustre and quality.

Adamite
Mapimi, Durango (Mexico)
Size: 8 cm × 8 cm

Adamite on limonite
Mapimi, Durango (Mexico)
Size: 12 cm × 8 cm × 6 cm

Smithsonite (Calamine)

The zinc deposits at Laurion in Greece were known to the Phoenicians over 3,000 years ago. The zinc ore known to ancient authors as cadmia was used in the manufacture of the copper-zinc alloy, brass. Cadmia was a collective name for carbonate and silicate zinc ores—smithsonite, hydrozincite, calamine and willemite. There is a genetic association of smithsonite-calamine so that they are found together in almost every zinc deposit. It was not until 1802 that an Englishman, J M L SMITHSON, the founder of the Smithsonian Institution in Washington, United States of America, succeeded in establishing the two as separate minerals. In 1832 zinc carbonate was named smithsonite in honour of the British mineralogist. In spite of its world-wide distribution, specimens showing good crystals remain a rarity. Smithsonite usually forms in botryoidal, reniform, stalactitic or encrusting masses. Matt to rough and curved faces are characteristic of the rhombohedral crystals.

The colour is very variable, and is determined by the inclusions of coloured substances, but greyish-white predominates. Yellow varieties of smithsonite result from included greenockite, brown from limonite, reddish-brown from hematite and green from copper or nickel minerals. The pink and lavender-coloured varieties are especially sought after.

The largest, finest and most vari-coloured smithsonite crystals have been found in Tsumeb and Abenab in Namibia. The strongly coloured reniform and botryoidal forms come from Magdalena County (New Mexico), Laurion in Greece, Iglesias on the Island of Sardinia, St Laurent du Minier in France and from Alston Moor in Cumbria, and Matlock in Derbyshire.

Pink smithsonite and an apple-green variety, that resembles chrysoprase, are used commercially in the decorative art and gemstone trades, though the low hardness severely restricts its use.

Smithsonite
Magdalena, New Mexico (U.S.A.)
Size: 11cm × 20cm × 7cm

Hydrozincite

Hydrozincite closely resembles smithsonite both in form and in composition. It occurs as a result of the weathering of zinc blende. Both smithsonite and hydrozincite are zinc carbonates; they differ only in that hydrozincite is a hydrated carbonate. Hydrozincite occurs in a variety of forms—fibrous, encrusting and massive, as small stalactites and fine, acicular rosettes. The colour varies from snow-white to pale yellow.

The occurrence of hydrozincite is the same as for smithsonite. Both minerals often occur in zinc ore deposits. Pure hydrozincite occurs only occasionally—near Santander in Spain and in Laurion in Greece. It also occurs at Llanidloes, Montgomery, Wales.

Smithsonite
Laurion (Greece)
Detail

Hydrozincite
Santander, Cantabria (Spain)
Size: 15 cm × 25 cm × 8 cm

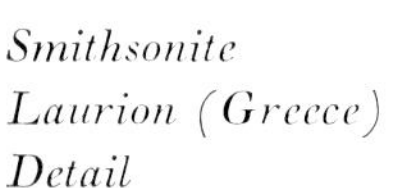

Minerals of Bismuth-Cobalt-Nickel Ore Deposits

The essential elements contained in the ores of this formation are silver, uranium, bismuth, cobalt and nickel. They occur in association with a carbonate-quartz gangue. When mining began in the Erzgebirge, silver and its ores were mined, and later nickel and cobalt ores were extracted, together with bismuth and pitchblende, from deeper levels. An abundance of different minerals is characteristic of these mineral veins. Besides silver, there are also sulphide silver ores containing minerals such as argentite, proustite, pyrargyrite and stephanite, while bismuth is found in association with skutterudite, chloanthite, pitchblende and the secondary uranium minerals. Nickeline and arsenic as well as nickel bloom (annabergite) and cobalt bloom (erythrite) also belong to this paragenesis.

The gangue minerals include calcite, siderite, witherite, dolomite, baryte, quartz and fluorite.

Bismuth

The longest known and most productive deposits of bismuth are in the German Democratic Republic in the Erzgebirge. Mention of the St Georg bismuth mine in the Wiesen (German for 'meadows') near Schneeberg, appears as early as 1472. The name of the metal probably derives from the place name. In German mining terminology the technical term 'muten' means to apply for a mining concession. Since the Georg mine—cited by G Agricola as the richest bismuth deposit—lay 'in der Wiesen' (in the meadows), 'Wis-mut' (German for bismuth) developed from 'mining in the meadow.' This derivation is admittedly uncertain, yet knowledge of the white lustrous metal came first from the Erzgebirge.

Proustite with skutterudite, chloanthite,
silver, arsenic, siderite and ankerite
Schneeberg, Erzgebirge (G.D.R.)
Size: 3 cm × 10 cm × 1 cm

The distinguishing features of bismuth are
such as to make it virtually unmistakable.
The silver-white colour of a freshly broken
surface rapidly tarnishes to a yellowish-red
tint. The high density (9.8) makes even hand
specimens feel quite heavy. The low hardness of bismuth is striking—it can be cut
without difficulty—and it has a low melting
point (271 °C). Wood's metal, a bismuth-
lead-tin-cadmium alloy, melts at 60 °C and
is used in the manufacture of trick objects,
such as spoons which melt in boiling water.

Although bismuth occurs in coarse, massive and foliaceous forms, until now there
has been little demand for it either from industry or from collectors. Arborescent and
reticulated forms are associated particularly
with the old mining town of Schneeberg in
the German Democratic Republic.

Bismuth crystals found in 1967 in the
Schneeberg mining field aroused the interest
of collectors, and have so far remained unsurpassed by finds from any other deposit.
The skeletal crystal faces are matt and coloured grey or violet-red from a coating of
skutterudite, while the cobalt changes by
oxidation to crusts of the crystals of cobalt
bloom (erythrite). The bismuth crystals,
which can be as large as 2 cm, have a metallic
lustre and tarnish only on broken surfaces,
and are frequently linked by skeletal crystals of bismuth and skutterudite on a matrix
of dolomite. Specimens with bismuth crystals
have a scarcity value, although specimens
with intertwined, dendritic or arborescent
forms on dolomite, the most attractive pieces,
show clearly the details of their structure
when polished.

Bismuth was not recognized as an element
until 1739, and prior to that was frequently
mistaken for antimony and zinc. It is now
used in the production of alloys and pig-
ments, as well as for cosmetic and medicinal
preparations. There are also deposits in Bolivia (Tasma, Oruro, Chorolque, Unica) and in
Australia (Kingsgate, Chillagoo), and it occurs in some of the mineral veins of Devon
and Cornwall.

121

Arsenic

Golden-yellow orpiment (arsenic trisulphide) was known before arsenic itself. This is the more remarkable, since native arsenic occurs as a mineral and was known in medieval times. Arsenic occurs typically as scaly encrustations and the curved fragments broken from stalactitic forms, the 'sherds,' were filled with water to kill flies. AGWERNER first introduced the name 'arsenic.'

Arsenic occurs in many ore veins in the Erzgebirge in the German Democratic Republic in association with silver and cobalt ores. The arsenic is usually dark grey to black, and only freshly broken surfaces have a silvery, metallic lustre, which soon darken and tarnish. The finest specimens with a dish-like form come from Pöhla, but are also found near Schneeberg. Arsenic is quite unmistakable when struck, for it emits a smell of garlic.

Globular masses of small rhombohedra are found at Akatani, in Japan, and it occurs in Arizona in the United States of America, and in Montreal and the Queen Charlotte Islands in Canada.

Native arsenic is of little commercial importance, since more arsenic than can be used is produced as a result of smelting ores of other metals which also contain arsenic.

Arsenic is used in the manufacture of lead-shot and insecticides, and was formerly used in some pharmaceutical preparations.

Proustite

The silver sulphide minerals range in colour from grey to black and are not particularly attractive, except for proustite which has a cherry-red to scarlet colour. AGWERNER differentiated between several ruby silver ores, distinguishing a light or pale ruby silver proustite, and a dark ruby silver ore—pyrargyrite—which contains antimony. The surfaces of both proustite and pyrargyrite tarnish on exposure to bright light by reason of the silver content, thus obscuring the translucent, deep red colour.

Erythrite on quartz
Neustädtel-Schneeberg, Erzgebirge (G.D.R.)
Size: 19cm × 9cm × 15cm,

The best known deposits are in the Erzgebirge in the German Democratic Republic and in Chile, where it occurs as very large crystals in parallel growth with smaller crystals. Fine crystallized specimens were obtained from the Schneeberg ore field, between 1955 and 1960. Scalenohedral crystals of up to 8 cm in length were found in association with arsenic, skutterudite and dolomite. Considerably smaller, but well crystallized specimens can be seen in collections from the Freiberg ore field. Weathering of marcasite and pyrite causes the specimens to deteriorate, and therefore proof of the richness of these rare silver ores is steadily diminishing.

Despite its high silver content, proustite has never been of industrial importance because of its rarity. It now finds a use in laser optics but the need for the crystals to be very pure and free from defects has greatly restricted its use.

Smaltite-Skutterudite-Chloanthite

The commonest cobalt and nickel minerals are members of the skutterudite series. They have almost identical crystal structures and because the cobalt and nickel cations are closely similar in size, they can replace one another in the structure and form mixed crystals. Since the bivalent (ferrous) iron cation is also closely similar in size to those of nickel and cobalt, it can also replace the other two elements. Smaltite is the cobalt-rich end-member of the series, and skutterudite contains appreciable amounts of both cobalt and nickel. Skutterudite was first described from Norway in 1827 by F.A BREITHAUPT. Slightly weathered specimens are the easiest to identify, because lustrous cubic crystals acquire a thin pink to violet coating of cobalt bloom (erythrite).

The finest and largest skutterudite crystals come from Schneeberg in the Erzgebirge,

Siderite with pyrite
Lobenstein, Thuringia (G.D.R.)
Size: 11 cm × 10 cm × 5.5 cm

Magnetite, mesitine spar with dolomite and quartz
Traversella, Piemont (Italy)
Size: 18 cm × 14 cm × 6 cm

German Democratic Republic, and from Bou Azzer in Morocco, and it also occurs in Cornwall.

Chloanthite-Skutterudite-Smaltite

Chloanthite is the nickel end-member of the smaltite-skutterudite-chloanthite series. On weathering, it takes on a green colour as a result of surface alteration to nickel bloom (annabergite), and the name derives from the Greek *chloanthos*, meaning 'green glimmer.' The difference in colour between weathered chloanthite and weathered skutterudite is an important diagnostic character, since it is very difficult to distinguish between the unaltered minerals simply on physical characteristics of colour, hardness, density and lustre. The curved and distorted faces of cubic crystals are often characteristic of chloanthite. Some of the reniform, botryoidal and fibrous intergrown forms of nickel-cobalt-iron arsenides have been shown to be safflorite. The occurrence of chloanthite is the same as for skutterudite and smaltite.

Erythrite (Cobalt Bloom)

Erythrite or cobalt bloom is a peach-blossom red encrustation formed by the surface oxidation of cobalt arsenide minerals (smaltite, skutterudite, cobaltite). The name, erythrite, is derived from the Greek word *erythros*, meaning red. The mineral mostly takes the form of small spherical, reniform, or radiating structures on the weathered surface or embedded in hollows of the rough surface of the ore. The colour of the encrustation varies with the humidity in the air; the deep pinkish-red fading to a pale bluish-pink under dry conditions. Good, acicular crystals, often with a star or cluster arrangement, are rarities. Prize specimens are known from the Schneeberg deposits in the Erzgebirge of the German Democratic Republic and Bou Azzer in Morocco.

Arborescent bismuth (feather bismuth)
on skutterudite with quartz
Schneeberg, Erzgebirge (G.D.R.)
Size: 5.5 cm × 9 cm × 3 cm

*Bismuth on dolomite, skutterudite and cobalt
bloom*
Schneeberg, Erzgebirge (G.D.R.)
Size: 13 cm × 11 cm × 8 cm

Magnificent crystal groups, usually on quartz, with deep red, radiating crystals up to one centimetre long were found in the Schneeberg ore field in the German Democratic Republic and until twenty years ago they were the best specimens known. Today, however, the best quality specimens come from the deposits at Bou Azzer. First-size specimens with tabular crystals up to 1.5 cm long in elongated cavities are not unusual. The intense, deep pinkish-red colour and the pearly lustre of the crystals are the chief features of Moroccan erythrite. Delicate coatings of small, acicular crystals only millimetres long from Mount Cobalt near Selwyn in Queensland (Australia) are likewise of interest to the collector, and specimens have become highly sought-after collectors' items within a short time. The drawback with these beautiful minerals is that the colour fades very quickly in water containing carbonic acid, the surfaces of the crystals then become rough and dull and take on a greenish-blue tinge. In England fine specimens of erythrite have been found at Alston Moor in Cumberland and at a number of localities in Cornwall.

Witherite

Witherite is a not particularly common barium mineral, which is usually found in veins in association with galena. A WITHERING, an eighteenth century English physician and mineralogist first drew attention to the mineral. It was named witherite by A G WERNER in honour of its discoverer. Single crystals of witherite are extremely rare. Repetitive twinning produces pseudo-hexagonal forms, and crystals resemble those of quartz. Witherite also occurs as reniform, botryoidal, fibrous and granular masses, occasionally having sheaf-like, hexagonal forms. It is milky-white to grey in colour, usually with matt crystal faces. Witherite can be distinguished from other carbonates by dissolving it in hydrochloric acid, and using the solution for the flame test, when the barium colours the flame green.

The finest specimens were obtained from Fallowfield in Northumberland and near Alston Moor in Cumbria; it also occurs in Durham. Specimens recently found at Schneeberg in the German Democratic Republic with reniform, botryoidal and sheaf-like formations rival those from England both in beauty and in size.

Witherite with pyrite
Schneeberg, Erzgebirge (G.D.R.)
Size: 14cm × 17cm × 8cm

Chloanthite with skutterudite, and nickel bloom and cobalt bloom
Raschau, Erzgebirge (G.D.R.)
Size: 9cm × 8cm × 2.5cm

Autunite, torbernite and uranocircite are, respectively, hydrated calcium, copper and barium uranium phosphates which are among the commonest and most important members of the group sometimes called the uranium micas, which occur in the oxidized zone of uranium ore deposits. They have a thin tabular, micaceous or scaly habit, and strongly resemble mica, although they are very different in colour. The uranium micas are much more brittle than the true silicate micas which are often, and mistakenly, taken thereby to be harder than the uranium minerals. The two groups of minerals are readily distinguished chemically: the uranium micas are predominantly phosphates and arsenates, whereas the 'normal' micas are silicates. The perfect cleavage, the property typical of the micas, is common to both groups of minerals.

Autunite is named after a locality near the town of Autun in the Saône et Loire département of France. It is a distinctive lemon-yellow colour, and occurs as dendritic forms in fissures in granite. The crystals occur mostly in groups or clusters. The yellow colour of autunite sometimes alters to pea-green, and it is strongly fluorescent in ultra-violet light. Localities in the Erzgebirge and Vogtland have yielded magnificent specimens. Also, it is found at Redruth and St Austell in Cornwall, the Flinders Range in South Australia, and at Peveragno, Piedmont, Italy. The best specimens are found at Streuberg near Bergen and at Zobes in Vogtland and are comparable with the large numbers of excellent specimens from Mount Spokane in the State of Washington in the United States of America.

Arsenic with proustite
Pöhla, Erzgebirge (G.D.R.)
Size: 20cm × 8cm × 12cm

Minerals from Other Hydrothermal Deposits

Mineral deposits of copper, gold, antimony, pyrrhotine, hematite and the tetrahedrite group of minerals are formed under conditions similar to those of the lead-zinc ore deposits and of nickel and cobalt, namely from hot (hydrothermal) solutions. The manganese ores, the copper ores, azurite, malachite and pseudomalachite and limonite occur only in the oxidized zones of such deposits.

Copper with a platy form
Lake Superior, Michigan (U.S.A.)
Size: 9cm × 15cm

Copper
Lake Superior, Michigan (U.S.A.)
Size: 32cm × 18cm × 8cm

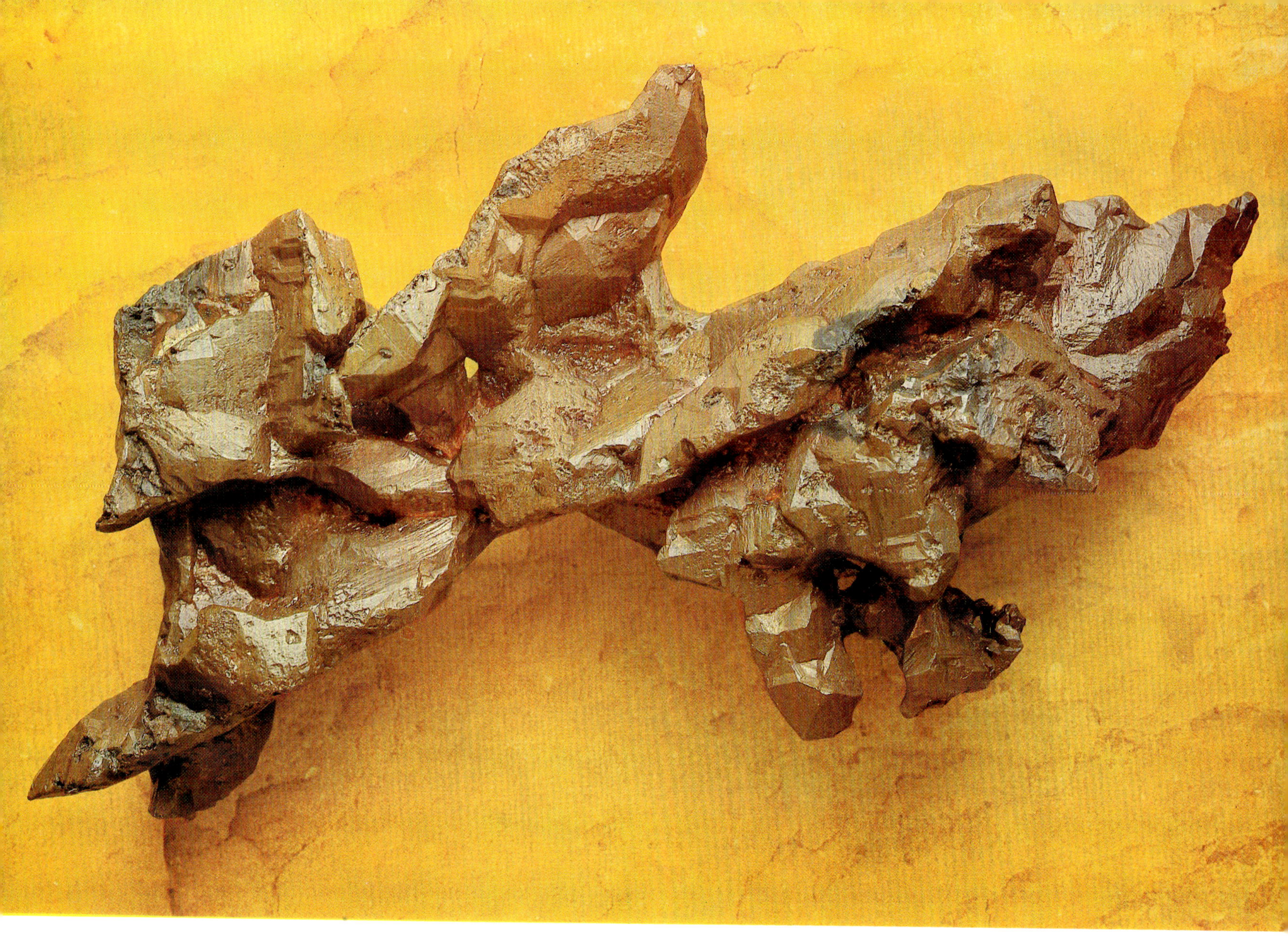

Copper

The origin of the name goes back to antiquity. Copper, together with gold and silver, is one of the oldest known metals. The Latin term *aes* acquired the meaning 'ore' in the Germanic languages, and was used also for bronze, the most important copper-tin alloy. It was at this time that the reddish colour of copper came to be incorporated into the name of the metal, for the red ore—*aes rubrum*—was also known as *cyprium aes* because its first known occurrence was on the Island of Cyprus. Literary references to copper ore from this Mediterranean island were abbreviated to *cyprum* or *cuprum*, from which in time the word 'copper' was derived.

Other literary sources claim that copper was first found in Chalcis on the Greek island of Euboea in the Aegean. It is from this that the names of several copper ores, such as chalcopyrite, chalcosine and chalcanthite are derived. G. AGRICOLA in his *Bermannus* gave a description of copper from Jáchymov (Joachimsthal) in Czechoslovakia, and likened the dendritic form of the copper in the mine-workings to unique works of art. Also famous is the sheet copper with faces as large as 1 sq metre, which were found near Zwickau (German Democratic Republic).

Copper typically crystallizes in cubes, which are often distorted; more rarely as octahedra and dodecahedra. The colour, low hardness, good malleability and electrical and thermal conductivity are further distinguishing features of copper.

The richest copper deposits occur near Bogoslovsk in the Urals (Soviet Union) and in the Keweenaw Peninsula (United States of America) on the southern shore of Lake Superior where, in 1869, 750 tons of copper were mined. Rich copper deposits occur also in Chile, near Andacollo and near Corocoro in the Atacama desert, in Bolivia. It has also been found at a number of localities in Cornwall.

Azurite (Blue Carbonate of Copper)

The cornflower blue of azurite makes it a particular favourite among minerals. From antiquity it has shared its popularity with lapis lazuli (lazurite), for translations of Arabic books on minerals often confused azurite with lazurite. Both names have the same origin, since the 'l' of lazurite was simply the Arabic form of the article.

The similarity in composition to malachite means that these two copper minerals frequently occur in association.

Malachite is the more stable form of the two copper carbonates, and there are many examples of the alteration of azurite into malachite. Distinguishing between the two minerals presents no difficulty. They differ in colour and their form and habit are also distinctive. Azurite does not occur as banded, agate-like formations, but usually forms prismatic, columnar crystals, which are frequently arranged in aggregates. Specimens of azurite and malachite from Namibia are much sought after, especially the pseudomorphs of azurite by malachite.

Pseudomorphs can be important, especially as a source of information about the conditions under which the minerals were formed. The azurite crystals have altered to malachite but have retained their original crystal form. Pseudomorphism can also result from the solution and transport of a mineral substance, so that a new mineral is formed by the crystallization on the original form.

Azurite crystals from Chessy near Lyons in France are well known. The deep blue, spherical aggregates resemble chestnuts in size and shape, but with a surface more like that of a pine-apple. Specimens from Bisbee, Arizona (United States of America) are remarkable in that deep blue azurite is tipped with pale green malachite. Specimens are also found in Cornwall associated with malachite.

Copper with cuprite
Rudabánya (Hungary)
Size: 12 cm × 11 cm × 5 cm

Azurite (one of the largest and finest
specimens from this deposit, with typical
rosette formation)
Rudabánya (Hungary)
Size: 10 cm × 10 cm × 6 cm

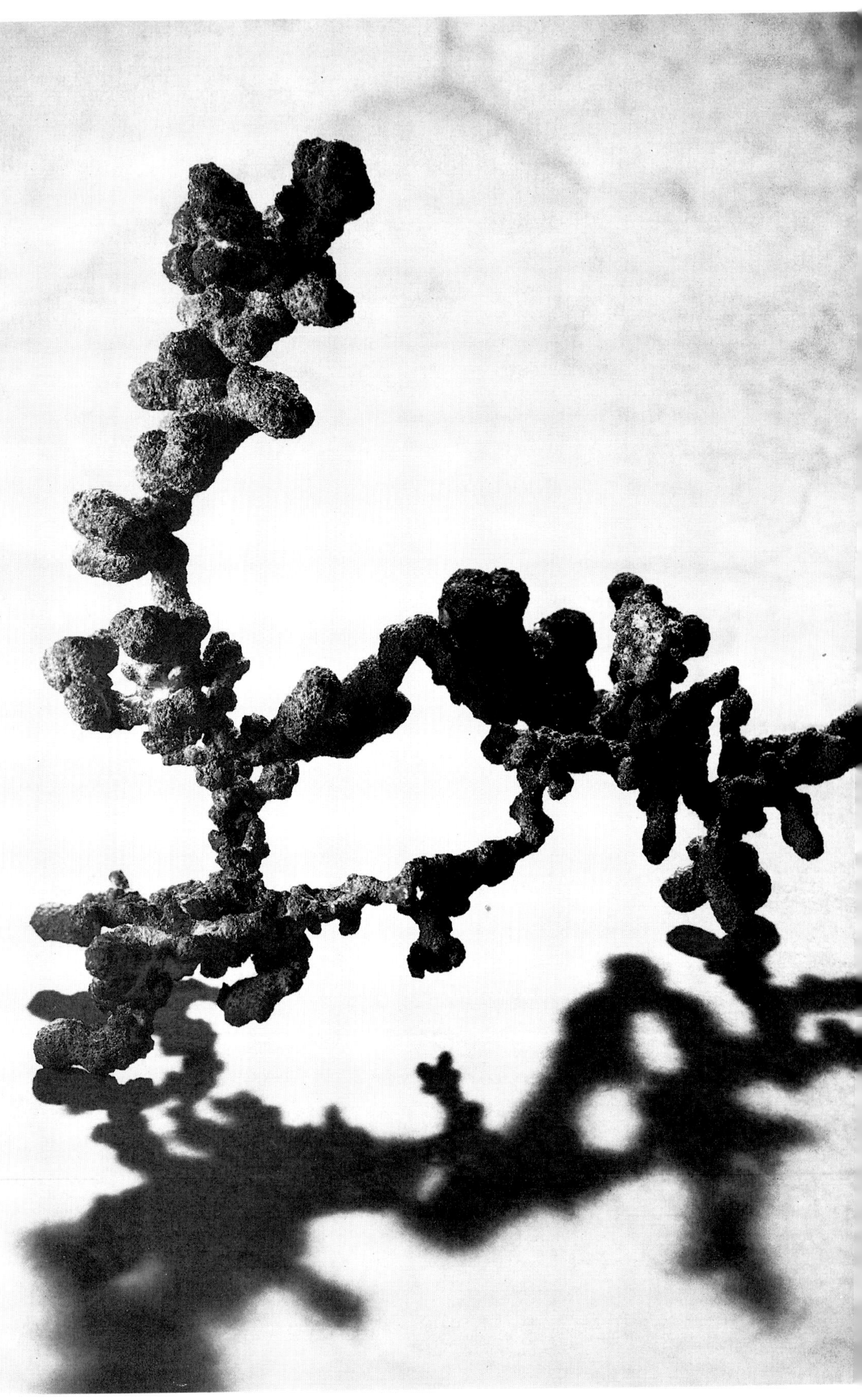

Malachite

Copper sulphide ores weather to produce
copper sulphate which alters in turn to malachite by interaction with carbonates. Malachite typically occurs as banded, layered masses which were deposited in cavities in association with limonite. Rare bright green
crystals have a silky lustre and are sometimes
translucent, and the contrast with the dark
background on which they rest is most striking. The rich green colour of malachite is so
characteristic that it was chosen as a colour
standard. The name, malachite, derives from
the leek-green colour of mallow (*malache* in
Greek). The Arabs attributed to it the power
of warding off fear and terror.

Apart from the colour, another distinguishing feature of malachite is its low hardness. It occurs commonly as massive forms
in copper ore deposits, and is consequently
very suitable for working. The spherical,
reniform or botryoidal forms are built up of
concentric layers, and the alternation of
light and dark bands is extremely decorative.
Malachite nodules with precisely concentric
banding arranged around a central core are
strongly reminiscent of agate. The difference
is that malachite grows from the centre out-
wards, while the opposite is true of agate.

The finest and largest pieces of malachite
suitable for working were found in mines in
the Urals, in the Soviet Union. Factories at
Sverdlovsk used the malachite to manufacture bowls, vases, platters and other objets
d'art. The use of malachite as a magnificent
decorative stone can still be seen today in
Leningrad in the Isaac's Cathedral, in the
Hermitage Museum, and in the Malachite

Azurite with malachite on limonite
Chillagoe, Queensland (Australia)
Size: 12 cm × 6 cm × 6 cm

Malachite (cut surface)
Mednorudyansk, Urals (U.S.S.R.)
Detail

Malachite, pseudomorph of azurite with
cerussite and bayldonite
Tsumeb (Namibia)
Size: 6 cm × 9 cm × 8 cm,
crystal size: 4.5 cm × 6 cm × 2 cm

Room of the Winter Palace. In 1836, a block
of malachite weighing more than 25 tons
and measuring 5 metres by 2.5 metres by
1 metre was extracted near Nizhni Tagil,
in the Urals (Soviet Union), and similar
specimens are known to have been found at
Gumeshevo. The world market is currently
supplied by Zaire, and by Australia, but
Cornwall has supplied some material, notably from Liskeard.

Pseudomalachite (Phosphochalcite)

There is very little difference in colour between the bright green of malachite and the
somewhat darker green of pseudomalachite.
The green colour of both minerals derives
from the copper content, yet they differ in
composition: malachite is a basic carbonate
and pseudomalachite a phosphate. The expert
can recognize the difference by the blackish-
green flecks on the greasy, vitreous surface
of pseudomalachite. The dominant form is
reniform to botryoidal, which, when broken,
has a fibrous, radiating structure. Stalactitic
forms are rarer and occur in cavities. Large
quantities of good pseudomalachite have been
found in the copper ore deposits of Nizhni
Tagil in the Urals (Soviet Union). The
specimens are unsurpassed by any other occurrences both in terms of colour and beauty
of form. Also worth mentioning are the surface encrustations or radiating clusters of
pseudomalachite on cellular limonite that
occur at Hirschberg in the Vogtland of the
German Democratic Republic, and in Virne-
burg near Mayen in the Rhineland of the
Federal Republic of Germany.

Stibnite on quartz
Oshio-yen, Iyo Province, Shikoku Island
(Japan)
Size: 10cm × 34cm × 7cm

Stibnite with quartz
Oshio-yen, Iyo Province, Shikoku Island
(Japan)
Size: 16cm × 20cm × 9cm

138

Gold

The articles of jewellery found in the tombs of the Pharaohs of ancient Egypt and the excavations of Troy, the fortified city on the Dardanelles in north-western Asia Minor, confirm that gold has been known and used since antiquity. It was then worked in its natural state, its extreme malleability being a great advantage. The jewellery, mostly chased—rings, bangles, clasps and necklaces as well as richly-decorated weapons—are evidence of the masterly skill of the craftsmen who made them. The reason why gold has always been so popular is without doubt its colour, and because, being a soft metal, it is so easily worked. Even J W von GOETHE, when describing the yellow colour, felt impelled to write of gold: "Gold in its altogether unalloyed state, especially if it also has lustre, gives us a new and intense concept of this colour." The Hebrew word for gold—*zahav*—means 'illumined by the sun's rays,' and the alchemists allotted to gold the symbol of the sun. The rapid change in colour of gold caused by admixtures of silver or copper recall Goethe's reference to its 'unalloyedness.' Pure gold contains $1-2\%$ silver, while the vast majority of natural occurrences have a content varying between $5-10\%$, but gold from the Altai and Transylvania is appreciably different with a silver content of between $25-30\%$. The pale yellow colour of this alloy, which is very similar to that of some varieties of amber, led to the use of the term, electrum (*electron*, derived from the Greek, means amber). Not only the colour is changed in gold alloys, but also the hardness and density. The less the amount of gold per unit measure, the greater the hardness of the alloy. This is particularly advantageous for articles of jewellery, which are consequently more resistant to wear. Workers in

Pseudomalachite
Nizhni Tagil, Urals (U.S.S.R.)
Size: 9cm × 8cm × 6cm

Pseudomalachite
Nizhni Tagil, Urals (U.S.S.R.)
Size: 9cm × 9cm × 4cm

mines and gold fields used to assess the approximate gold content of a find by the colour of the streak, produced when the specimen was rubbed gently on a dense, even-grained black touchstone.

The oldest and formerly most productive known gold deposits are those of secondary origin, the placer deposits. Continuous weathering of gold-bearing rocks by heat, frost and rain released the gold, and transported it in streams and torrents. As the velocity of the streams decreased, the high specific gravity of the gold (19.3) caused the metal to settle to the stream bed. Minerals with high specific gravities and which are stable under atmospheric conditions, for example, gemstones such as zircon, corundum, diamond and garnet, and precious metals such as gold and platinum, are frequently found as placer deposits in river beds and on the sea-shore.

The largest nugget of gold ever found in placer deposits came from Victoria in Australia and weighed 70.9 kilos, and one from Miass in the Urals (Soviet Union) weighed 36 kilos, the latter found in 1842. The wealth of gold in the Duero and Tajo rivers made Spain, while under Phoenician rule, the most coveted country in the world.

Although no commercial gold mining is now carried out in Britain, formerly it was important. The Romans mined gold in Wales, but it was during the second half of the nineteenth century that the Welsh mines flourished and in 1904 production reached 19,655 ounces. The wedding rings of the Queen and Princess Margaret are made from gold mined at the Clogau mine, Merionethshire.

Gold with quartz (144g, gift to the city of Freiberg, 1916)
Barberton District, Transvaal (South Africa)
Size: 8cm × 3cm × 3cm

Gold in leafy form on quartz
Musariu, Transylvania (Rumania)
Size: 12cm × 14cm × 4.5cm

Most of the Cornish stream gravels that have been worked for alluvial tin (cassiterite) have also yielded small quantities of gold. Gold, as well as silver, was obtained from the copper and lead ores of the Gold Scope mine in the Vale of Newlands in the Lake District during the reign of Queen Elizabeth I, and there are numerous other small, but uneconomic occurrences in Cumbria. Alluvial gold has been found at a number of Scottish localities including the area around Leadhills and Wanlockhead, and in Sutherland.

Gold occurs at Gastein in Salzburg (Austria), Beryesovsk near Sverdlovsk in the Urals (Soviet Union), Witwatersrand in South Africa (the largest deposit in the world), the western slopes of the Sierra Nevada in California, United States of America (Monumental Mine with one find of 43 kilos), Cripple Creek in Colorado, United States of America, the Klondike area and Bonanza

Stream in Alaska, United States of America, and Kalgoorlie in Australia. There are numerous other localities, though not many of these are of economic significance.

Gold is used—now as ever—in the manufacture of jewellery, for coins (mostly alloyed with copper, silver or nickel) and gold bars are the legal foundation of many currencies. It is also used in many ways in the china industry and in dentistry.

The annual world production figure for gold is in the region of 1,000 tons.

Pyrrhotine
Stari Trg, Kosmet, Serbia (Jugoslavia)
Size: 6.5 cm × 5.5 cm × 2 cm

Hematite, mammillated form (kidney-iron ore)
Ulverston, Lancashire (Great Britain)
Size: 10 cm × 10 cm × 6 cm

Hematite, showing growth form
Congonhas, Minas Gerais (Brazil)
Size: 9cm × 8cm × 3.5cm

Hematite (kidney ore)
Stafford (Great Britain)
Size: 10cm × 8cm × 9cm

145

Stibnite (Antimony Glance)

In ancient Egypt powdered stibnite was used in ointments for make-up or to darken the eyebrows and eyelashes. The Greeks also used it for eye disorders and called it stibium, hence the name, stibnite. Later, after 1000 A.D., the term *antimonium*, derived from the Arabic, came into use. In the Middle Ages, German miners named antimony ore spear glance or spear glass, after its characteristic columnar or acicular, lead to steel-grey, metallic crystals, which show iridescence on the vertically striated faces.

The finest crystal specimens formerly came from the Island of Shikoku in Japan. The large size of the crystals with their fabulous lustre make these specimens quite spectacular. Crystals 0.5 metres long and 5.5 cm across were found in ore veins in crystalline schists. Almost as striking are the specimens from the Hunan Province in China.

Specimens of stibnite from Baia Sprie in Rumania are especially prized. The long, acicular crystals are often penetrated by thin tabular baryte, which is either colourless or tinted red by realgar. This combination of minerals is particularly attractive. The lustrous, iridescent needles of stibnite from Kremnica in Czechoslovakia frequently form a felted mass of crystals, and, on account of the loose or thin matrix they are extremely brittle.

Pyrrhotine (Magnetic Pyrites, Pyrrhotite)

Together with pyrite and marcasite (FeS_2), pyrrhotine is another iron sulphide with the composition FeS. Whereas pyrite is usually a pale, light colour, the characteristic colour of pyrrhotine is a reddish bronze, mostly with golden brown iridescence. It was this feature that influenced the Greek choice of name, pyrrhotine (*pyrrhos* is Greek for fire-coloured, golden yellow). Its directional magnetism is a diagnostic feature: it is strong and permanent along the principal axis of the crystals, but at right angles it is barely perceptible. A G WERNER named the mineral magnetic pyrites for this reason. Well formed crystals occur in only a few deposits. By far the largest and most beautiful specimens were found in Serbia (Jugoslavia). Other occurrences are Herja in Rumania, Bottino in Tuscany (Italy) and Santa Eulalia in Mexico.

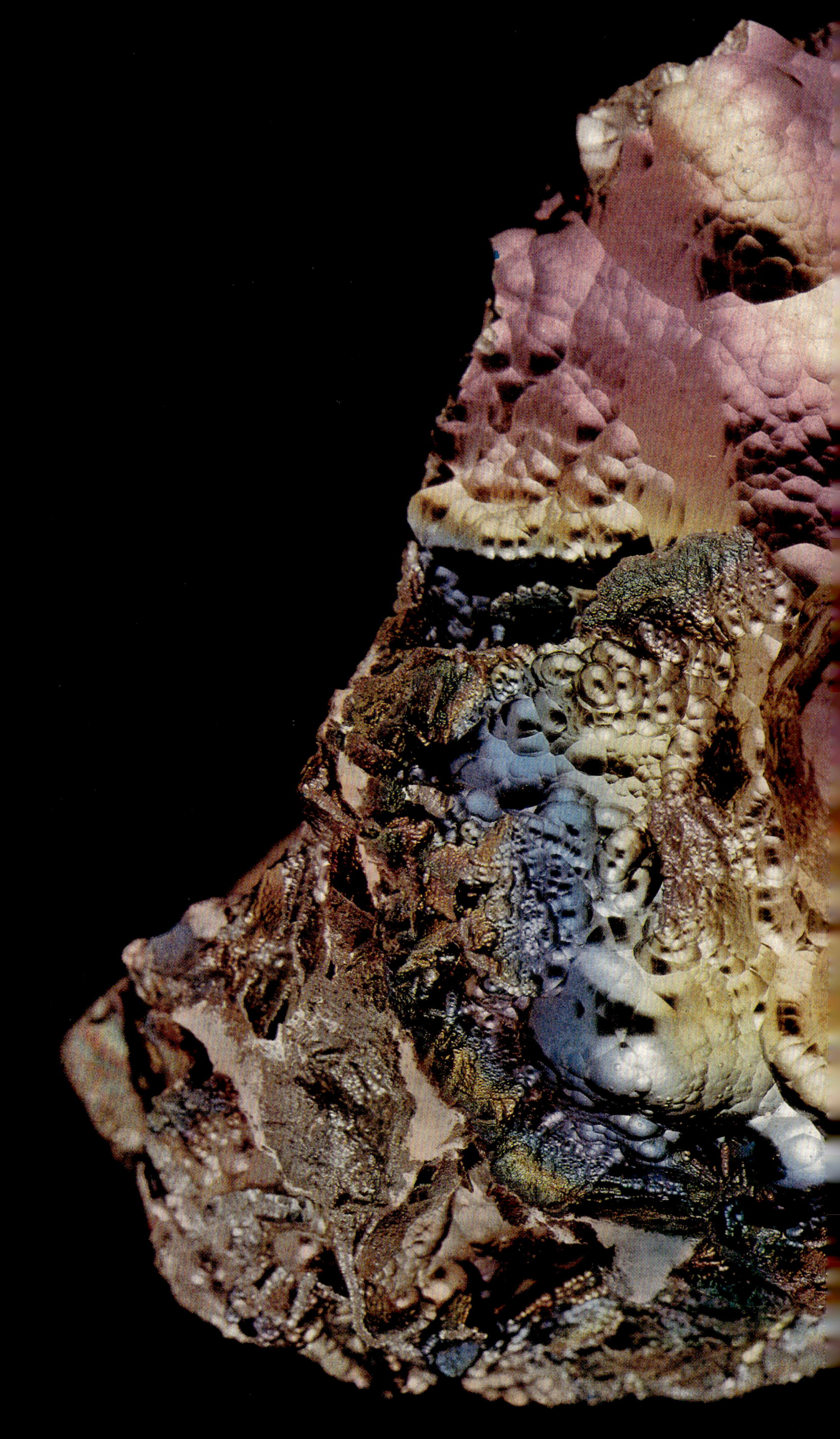

Hematite with annealing colour
Horhausen, Westphalia (Federal Republic of Germany)
Size: 10 cm × 7 cm × 5 cm

146

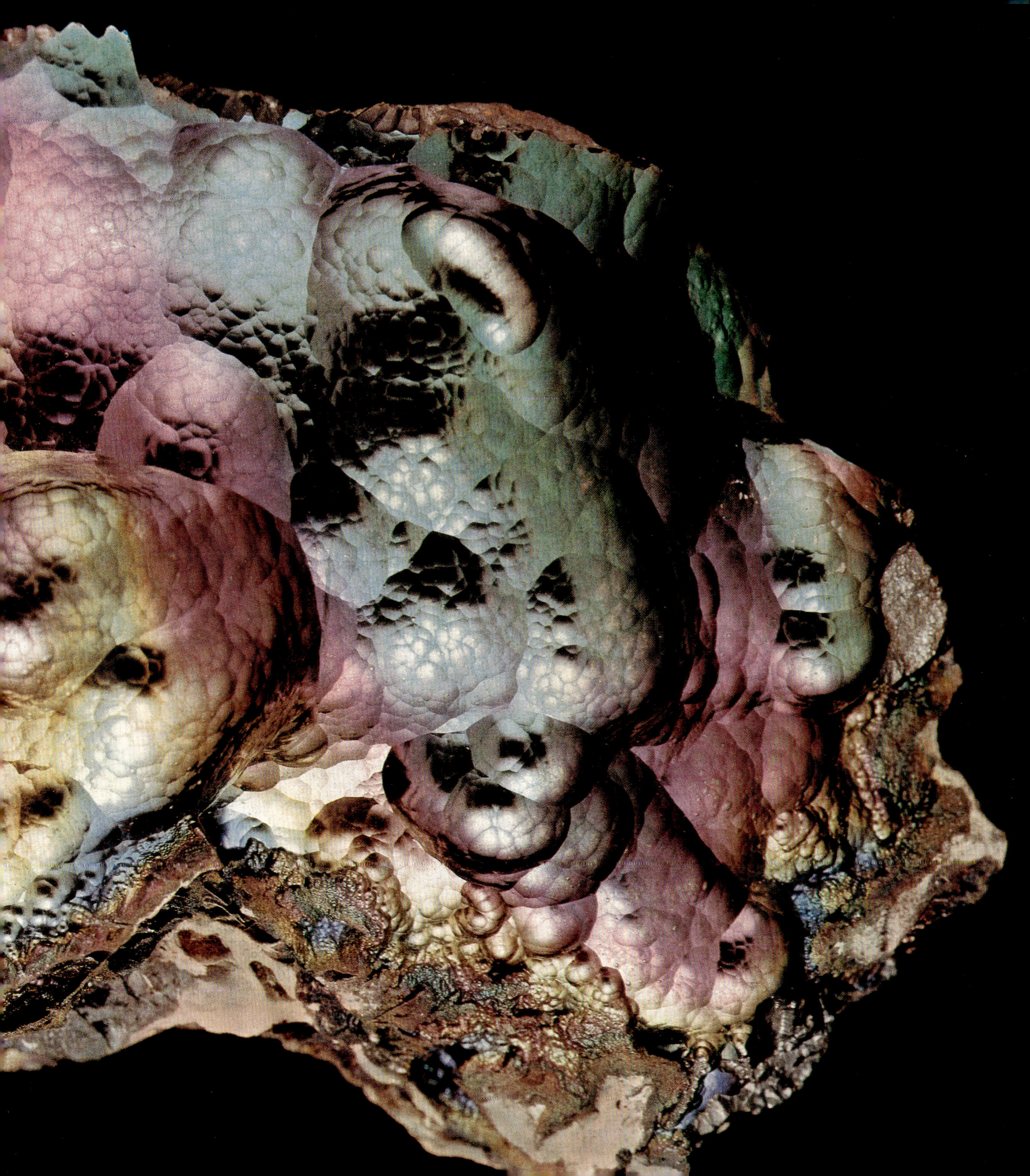

*Hematite (Red Iron Ore, Blood-
stone, Specular Iron Ore)*

This widely distributed iron ore is of great
interest on account of its variety of form and
habit and its colour, especially when the surface has a rainbow-coloured iridescence. The
colour varies according to the size of the
grains of the mineral and ranges from grey-
black to bright red. The colour of massive
hematite closely resembles that of blood and
in powdered form the likeness is even more
striking. The name hematite is derived from
the Greek word *haema* meaning 'blood.'

Hematite is an excellent example of a
mineral which crystallizes in different habits,
depending on the conditions in which it is
developed. With increasing temperature delicate, tabular forms give place to lenticular
to spherical habits, and finally to pyramidal
and even pseudocubic crystals.

The scaly, loose masses of hematite stain
deeply, and are widely known as 'reddle' or
'ruddle.' Radiating, botryoidal forms are
common and, because of the shapes produced,
are sometimes known as 'kidney ore.'

In ancient Egypt and Babylon, hematite
was put to a variety of uses. The dense variety
is easily ground and polished, and was worn
as jewellery. Its styptic (blood-staunching)
property was employed in medicine. An
early form, often mixed with clay, was used
for polishing precious metals and as a dye and
in many parts of the world it has been used
as a cosmetic.

Only a few of the many deposits known
can be mentioned here. Particularly fine specimens of mamillated hematite (kidney ore)
come from the mines near Ulverston in
Lancashire. Similar forms are known from
Zorge in the Harz of the Federal Republic of
Germany. Fine crystals come from the Island
of Elba, and from the volcanoes of Etna and
Vesuvius. Massive hematite is exploited as
iron ore at many places including the area of
the United States of America adjacent to
Lake Superior, the Ukraine, Soviet Union,
Australia and South Africa.

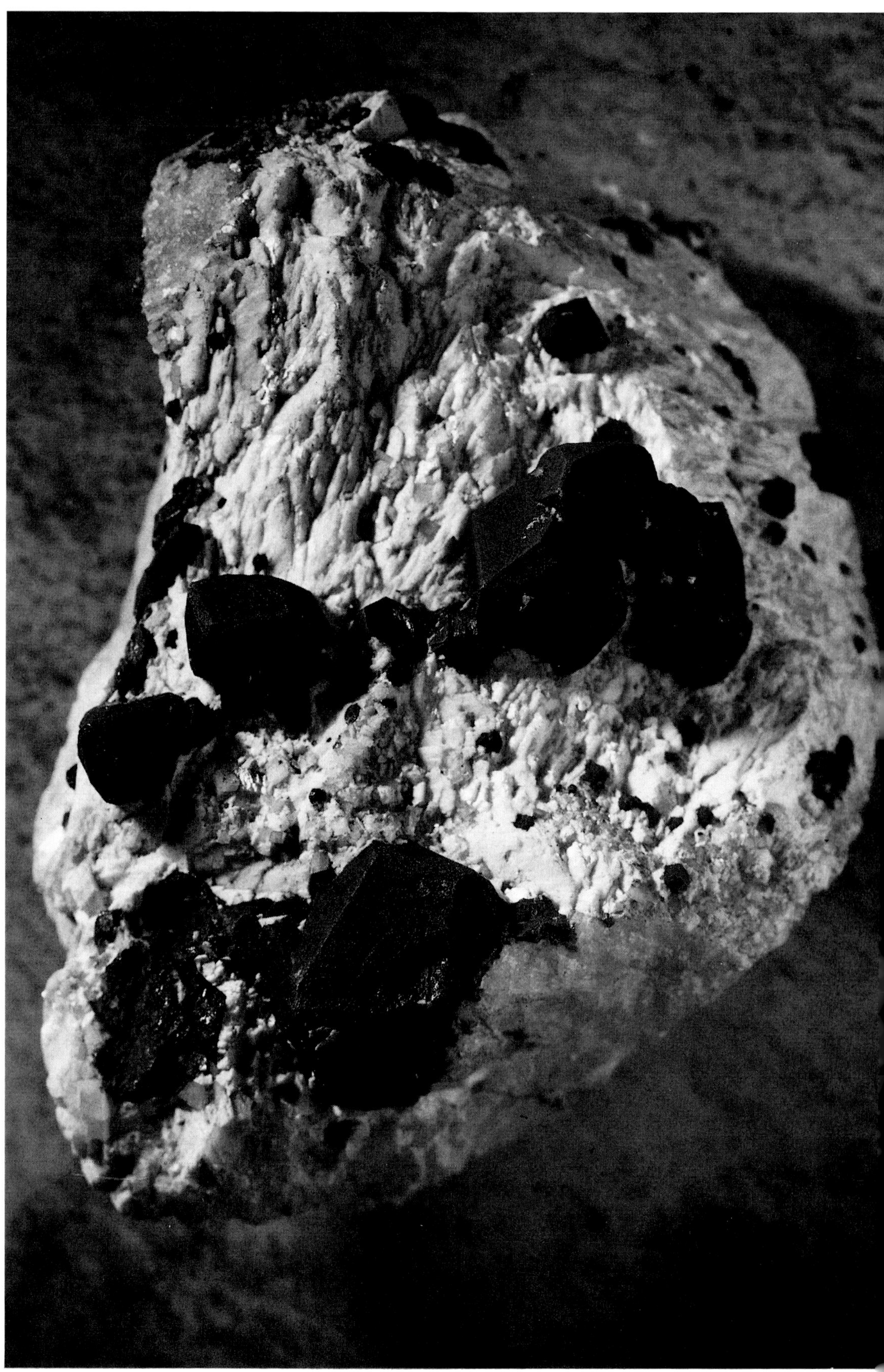

*Schwazite (variety of tetrahedrite) with
baryte on calcite and dolomite
Kogl near Schwaz, Tyrol (Austria)
Size: 10cm × 6cm × 3.5cm*

The most impressive and valuable forms of
hematite are the iron roses, which are specific to the Alps and are a form of specular
iron ore. This variety has an intense metallic
lustre, and the mirror-smooth crystal faces
sparkle with iridescent colours. Growth striations can also be detected on them, as slight,
but sharply bounded undulations, resembling
in some way the blocks of parquet flooring.
These highly-prized collectors' items occur
at Cavradi in the Tavetsch valley, St Gotthard region, the Binnatal in Switzerland, the
Island of Elba, the Italian volcanoes Vesuvius
and Stromboli, and Minas Gerais in Brazil.

Schwazite

Schwazite is a mercury-bearing variety of
tetrahedrite, which took its name from the
place where it was found—the Schwazer
Erzberg in the Tyrol, Austria. The iron-
black rhombdodecahedral crystals occur in
dolomitic limestone. The members of the
tetrahedrite group are copper minerals, in
which various elements (silver, zinc, iron,
mercury, bismuth) can replace part of the
copper in the structure. Members of this
group of sulphides while sharing the same
physical and crystallographic properties,
vary considerably, however, in their chemical
composition. The typical crystal form is the
tetrahedron, from which the name tetrahedrite is derived. The varieties containing
mercury and bismuth are known to occur in
only a few localities.

Limonite and Goethite

Many differing types and admixtures of iron-
bearing minerals of varying form and colour
are included under the general term limonite, and in the past were regarded as independent minerals. Much of what has been
described as 'limonite' has been shown to
consist of crystalline goethite. The origin
of the word, limonite, is not clear: the derivation from *limus*, the Latin word meaning
'slime' would correspond to bog-iron ore,
while tracing it back to the Greek *leimon*
(meadow), would make it equivalent to the

Limonite, Bilbao (Spain)
Detail

149

old name, meadow-ore. Both names give a useful indication as to the development of limonite, which is widely distributed and readily identifiable by its colour. Limonite forms by oxidation at the Earth's surface, or is precipitated by flocculation as mud near the banks of marshy ponds. Since this process can frequently be observed in acidic marshes, the names bog-iron ore and meadow-ore are very apt. Moreover, there is a considerable likelihood of foreign matter becoming enclosed during the flocculation of limonite. Under such conditions of formation some variation in the properties of limonite is only to be expected. Limonite can be loose and earthy, or massive and compact. The gel-like precipitates often have reniform, botryoidal forms with a fibrous or scaly structure, and stalactites, typical sinters and concretions frequently occur. The botryoidal forms with their smooth, nearly black and occasionally iridescent surfaces, can attain a considerable size. They have a silky to waxy lustre on the fibrous fracture surface.

The colour of limonite varies between yellowish brown, yellow ochre, and clove to dark brown, depending on the water content and the grain-size. Bog-iron, in particular, is a dark brown colour.

The streak is a diagnostic feature of the different iron and manganese ores. Streak is the colour of the powdered mineral and is produced by drawing it across an unglazed porcelain plate. The streak of hematite is reddish-brown, that of limonite is light brown to brownish yellow (brown iron ore), and that of manganese ores is black.

Limonite with radiolitic reniform (brown)
Rossbach near Siegen, Westphalia (Federal
Republic of Germany)
Size: 10cm × 8cm × 4cm

Mammillated limonite with radiolitic form
(brown)
Siebenhitz near Hof, Bavaria (Federal
Republic of Germany)
Size: 7cm × 5cm × 4.5cm

Pyrolusite on quartz porphyry
Jüchnitzgrund near Elgersburg, Thuringia
(G.D.R.)
Size: 10cm × 7cm × 4cm

Pyrolusite with psilomelane
Rossbach near Siegen, Westphalia (Federal
Republic of Germany)
Size: 13cm × 12cm × 5cm

The phosphatic limonitic ores of Lorraine
have become particularly important. When
mined by the Thomas process, these 'minette'
ores yield an important fertilizer for use in
agriculture. Limonite occurs worldwide and
in large quantities, and so is an important
iron ore. A few of the localities which are
represented in museum collections are Framp-
ton Cotterell near Bristol and Lanlivery,
St Just and other areas of Cornwall. Pseudomorphs of limonite after pyrite come from
the Island of Elba and there are a number of
localities in France, Czechoslovakia, Austria,
German Democratic Republic and Federal
Republic of Germany and Sweden.

Pyrolusite

The oxide ores of manganese have very similar compositions. The essential constituent is
manganese dioxide (MnO_2) with greater or
lesser amounts of impurities incorporated as
a gel during precipitation. It is these impurities that cause differences in the properties
of manganese ores and as a result many mineral names have been given to the individual
varieties.

Manganese ores are usually black or grey.
The brown glazes of early earthenware were
made with manganese ore.

Pyrolusite is one of the commonest and
most important manganese minerals. It was
used until the eighteenth century to remove
the colour from glass stained brown by iron
compounds. In order to improve its transparency, the molted glass was 'washed' by the
addition of pyrolusite; the name alludes to
this process, for *pyr* is Greek for 'fire,' and
luein means 'to wash.'

Pyrolusite occurs in a variety of forms,
radiating crystalline masses being the most
common. The thin radiating fibres occur in
clusters, whose metallic lustre and iron-black
to steel-grey colour give them a most attractive appearance. The low hardness and softness cause pyrolusite to soil the fingers when
handled.

Psilomelane
Schneeberg, Erzgebirge (G.D.R.)
Size: 14 cm × 11 cm × 8 cm

154

Well-formed, single crystals occur in only a few deposits. The intense grey metallic lustre of this variety of pyrolusite earned it the name, polianite, from the Greek word, *polios* meaning grey. Pyrolusite deposited as dendritic or arborescent forms on a pale background, as in the lithographic limestone from Solnhofen in the Federal Republic of Germany, has a striking appearance. Fine specimens of pyrolusite come from deposits in Elgersburg, Geraberg and Öhrenstock in Thuringia, Ilfeld in the Harz, German Democratic Republic, Horní Blatná in Bohemia (Czechoslovakia), and Lanlivery in Cornwall. Large deposits occur in the Central Provinces of India where it is associated with psilomelane.

The largest manganese deposits in the world are at Chiaturi in Georgia and Nikopol in the Ukraine (both in the Soviet Union).

Manganese is one of the metals of industrial importance which may eventually be extracted from ocean bottom sediments. Manganese nodules, which are abundant in parts of the Pacific and the Atlantic, contain pyrolusite and significant amounts of copper, zinc and other metals.

Psilomelane

Reniform and botryoidal forms are characteristic of this manganese ore. The surfaces can be either shiny black or have a dull submetallic lustre. The smooth, lacquer-like crusts of psilomelane nodules led to its name, which is derived from the Greek for *psilos*, meaning bald or smooth, and *melos*, meaning black. The scaly or shelled structure of the manganese ore is plainly visible on the conchoidal fracture surfaces, which also show inclusions of other substances.

Psilomelane with radiolitic form (black)
Langenberg near Schwarzenberg, Erz-
gebirge (G.D.R.)
Size: 18cm × 17cm × 9cm

Psilomelane (dendritic form on dolomite)
Solnhofen near Pappenheim, Bavaria
(Federal Republic of Germany)
Size: 15cm × 10cm

156

The type material of psilomelane comes from Schneeberg in the Erzgebirge of the German Democratic Republic. Until the eighteenth century, it was the commonest form of manganese ore found there, occurring in the iron lodes in association with hematite and limonite.

The finest examples of reniform and stalactitic psilomelane were quarried in the Schneeberg field. Psilomelane containing some lithium and copper occurs also in the bordering areas of Johanngeorgenstadt, Breitenbrunn, Eibenstock and Ehrenfriedersdorf, and was quarried there even earlier than at Schneeberg in the German Democratic Republic.

Other localities include Lanlivery in Cornwall and the Central Provinces of India. It also occurs in Minas Gerais, Brazil, and in the United States of America in association with the hematite deposits of the Lake Superior area.

Manganite

The finest and oldest finds of manganite come from Ilfeld in the Harz Mountains in the German Democratic Republic. The black, columnar crystals contrast vividly with the associated white baryte or calcite. The metallic lustre of manganite and the striations along the length of the prism faces are striking. Another helpful diagnostic feature is the radial arrangement of the crystals. The name refers to manganese, the main constituent of the mineral. In 1823, F A Breithaupt described the ore as *Glanzmanganerz*—lustrous manganese ore—thus emphasizing the dominant property of the mineral and its principle constituent.

It occurs at various localities in Cornwall, including the Botallack mine, St Just. Also from Exeter, Devon, Egremont in Cumbria, and from near Towie in Aberdeenshire.

Psilomelane, pseudomorph of limonite
Wissen an der Sieg, North Rhine-Westphalia (Federal Republic of Germany)
Size: 15 cm × 16 cm

Manganite
Ilfeld, Harz (G.D.R.)
Size: 9 cm × 8 cm × 6 cm

Pneumatolytic Minerals

As the residual melt of a magma cools, it becomes enriched in volatile components and eventually crystallizes. These processes occur at temperatures between 400–550 °C fairly close to the Earth's surface, and result in the formation of pneumatolytic minerals. Typical pneumatolytic minerals are wolframite, cassiterite, scheelite, molybdenite, apatite, fluorite, zinnwaldite, tourmaline and topaz, and several tin-tungsten ore deposits will be described in some detail.

Wolframite

Wolframite ore has been known since the sixteenth century. It was of no industrial importance at that time but being associated with cassiterite, was a by-product of tin mining, for during smelting, the tungsten carries some of the tin with it into the slag. The suffix *–ram* in the name 'Wolfram' means 'soot,' and 'Wolf-ram' really means 'loss of the tin'—for the 'soot' was thought to eat the tin. Further, the pitch-black wolframite could not be separated from cassiterite either by crushing or washing, because it has a high specific gravity, similar to that of cassiterite.

Chemically, wolframite is an iron manganese tungstate. The pure iron and the pure manganese tungstates are separate species called ferberite and hübnerite respectively. The colour varies according to the content of iron and manganese—iron-rich wolframites are dark brown, while an excess of manganese produces a deep black colour. Wedge-shaped crystals are typical, but thick, tabular, elongate and acicular forms also occur. Particularly fine crystals with a metallic lustre were found in the tungsten deposits of the Erzgebirge in the German Democratic Republic and Czechoslovakia (Cinovec and Horní Slavkov) and in Panasqueira (Portugal) and Chorolque in Bolivia. It is also found associated with tin ores at a number of Cornish localities, including St Austell.

Tourmaline, var. verdelite
Santa Rosa (Brazil)
Size: 4cm × 16cm × 4.5cm

158

Scheelite

Scheelite is an interesting mineral because of
its fluorescence: in ultra-violet light it fluo-
resces with a characteristic and intense bluish-
white colour.

Scheelite was first described by A F CRON-
STEDT in 1758; he called it *tungsten* in Swedish—meaning 'heavy stone'—and regarded
it as an iron ore. In 1781, K W SCHEELE discovered tungstic acid in *tungsten*, from which
the brothers F and J J DE ELHUYAR obtained
the metal, tungsten, two years later. Meanwhile, in honour of the discoverer, the name
'scheelite' was finally adopted.

Scheelite crystals are mostly found on
quartz, in association with zinnwaldite and
wolframite. They vary from whitish-grey to
brownish-yellow though colourless crystals of
gemstone quality were found in California,
United States of America. The predominant
crystal form is the tetragonal bipyramid. It is
found at Carrock Fell in Cumbria and associated with the tin veins of Cornwall. Also
at Gyojomen, Chushu-gun, Chusei-Hoku-
Do in Korea. Besides these, there are deposits in Horní Slavkov in Bohemia (Czechoslovakia), Traversella in Italy, as well as
crystals weighing one hundredweight and
more from Omaruru in Namibia and from
Brazil.

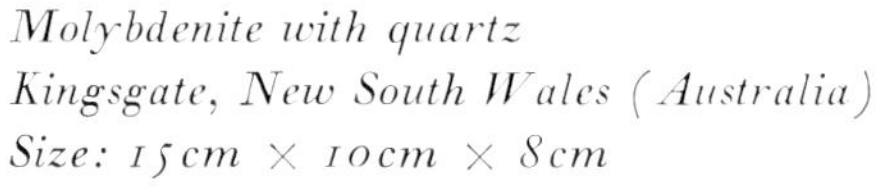

Molybdenite with quartz
Kingsgate, New South Wales (Australia)
Size: 15cm × 10cm × 8cm

Zinnwaldite
Zinnwald, Erzgebirge (G.D.R.)
Size: 8cm × 9cm × 6cm

Molybdenite

The Greek word *molybdos* (meaning lead) was applied originally to lead, lead ores and artificial lead products. It was applied to the mineral now known as molybdenite (determined by D L G Karsten in 1908) because it was confused with massive, fine-grained galena and graphite. The distinguishing features of molybdenite are its bright metallic lustre, lead-grey to bluish grey colour, flexibility, perfect cleavage and low hardness.

Well-formed crystals are rare; molybdenite occurs mainly as platy or scaly aggregates or rosettes. The six-sided, tabular crystals frequently lack regular outlines. Bent and deformed platy crystals frequently occur in granite pegmatites.

Distinguishing between molybdenite and graphite is relatively simple. The dark grey streak of molybdenite takes on a dirty green tone when rubbed hard, while that of graphite does not alter.

Molybdenite is the most important ore of molybdenum and, although it occurs widely, it is rarely found in large quantities. It is usually associated with deposits of scheelite, wolframite, cassiterite and fluorite. It occurs at Carrock Fell in Cumbria, but the best crystals come from Edison, New Jersey, United States of America, Wakefield, Quebec and Belleville, Ontario, Canada and from New South Wales, Australia. The most extensive ore deposits are in Colorado, United States of America and in Telemark in Norway.

Zinnwaldite

Zinnwaldite is a member of the mica group of minerals, and mica was already known to miners in the sixteenth century. It aroused interest on account of its lustre, but was despised since metal could not be extracted from it. The contempt of the mineworkers was expressed in the term *Katzensilber*, German for 'cat silver,' the silver referring to the lustrous surface, and cat to the absence of useful metal.

Scheelite with zinnwaldite on quartz
Zinnwald, Erzgebirge (G.D.R.)
Size: 9cm × 6cm × 5cm

Vanadinite
Mibladen (Morocco)
Size: 2.5cm × 2.5cm × 1.5cm

Vanadinite
Mibladen (Morocco)
Size: 8cm × 1.5cm × 3cm

All micas have a perfect cleavage, so that
wafer-thin flakes can be detached with ease
from the crystals. They are usually classified
into light and dark micas. Muscovite is the
best known and commonest of the light micas,
and iron-bearing biotite is a reddish-brown
colour and is the most widespread of the dark
micas. Zinnwaldite varies in colour from sil-
ver-grey to brown, through green, and pale
violet to almost black. The name is derived
from the Zinnwald tin deposits in the Erzgebirge in the German Democratic Republic,
where there are crystals of zinnwaldite several centimetres in size associated with
smoky quartz, cassiterite, scheelite, wolframite and fluorite. The thin tabular, six-sided
crystals often occur in fan-shaped groups.
The lithium in zinnwaldite colours the flame
red. Other important localities for zinnwaldite
are the St Just area and elsewhere in Cornwall, Narssalik in Greenland, and the Seward
Peninsula of Alaska.

Tourmaline

The tourmaline group of minerals is unri-
valled for its rich colour range. Tourmaline
is a complex borosilicate containing more than
a dozen elements which are responsible for
the diversity of colour shown by different
varieties. Iron, chromium, vanadium and
manganese have a particular influence on the
colour. In 1703 seamen sailing to the Dutch
East Indies brought red and brown tourmaline back to Europe; the Singhalese called
these gemstones *turamali*. In its new-found
home, tourmaline was not prized as a gemstone, but for its property of acquiring an
electrical charge when rubbed or warmed.
The seamen used their tourmalines to clean
Meerschaum pipes: the electrically charged
crystals attracted the particles of ash, and
were known as 'Aschetrekker.' This name
considerably lowered the value of tourmalines, and consequently they were virtually forgotten.

Wolframite with quartz, pyrite, arseno-
pyrite, apatite
Beira Baixa (Portugal)
Size: 11cm × 9cm × 10cm

162

The prismatic crystals, usually with a tri-angular cross-section, vary in size; they commonly occur as needles, but single crystals can be up to one metre long and weigh as much as fifty kilograms. The faces have a vitreous lustre, while fracture surfaces are uneven and matt. Tourmaline shows a great range of colours, and colour-zoning is common. Tourmaline is a good example of a pleochroic mineral: polarized light is absorbed to a differing degree depending on the orientation of the crystal with respect to the polarization plane and, in consequence appears light and dark.

Schorl is iron-rich and is the commonest variety of tourmaline. Fine black crystals are found in the area of Bovey Tracey in Devon, and it occurs at many localities in Cornwall associated with the granites. The extraordinary rock luxullianite consists essentially of radiating clusters of tourmaline needles and quartz and is named after Luxullyan, in Cornwall.

Rubellite is pink to pinkish violet in colour and also occurs as crystals having two colour zoning in red and green. It occurs in the Urals, on the Island of Elba, in San Diego County, California, in Madagascar, in Mozambique and in Brazil.

Indicolite is a rare blue variety of tourmaline found in the Urals, Namibia and Brazil.

Dravite is brownish-black and occurs in the Drau region of Carinthia and the Zillertal in the Tyrol (Austria).

Achroite is colourless or can have a faint green tinge. Crystals from the Island of Elba have black tips and these are known in German as 'Mohrenkopf' (Moor's head).

Verdelite is green, and occurs in the Urals associated with chromite deposits. The other main source is Brazil.

Halite (Rock-salt)
Wieliczka near Cracow (Poland)
Size: 15 cm × 15 cm × 12 cm

164

Minerals of Sedimentary Origin

Weathering is a process producing changes in minerals and results from forces acting at the Earth's surface. These include water, air, wind, heat and cold, as well as all kinds of organisms. Under their influence rocks are broken down; one part remains as a weathering residue, or is transported as solid particles, the rest is dissolved and transported elsewhere by water. After transportation the dissolved substances are once again precipitated as new minerals. Salt deposits are formed in this way, as a result of evaporation, the commonest of such minerals being rock salt (halite), anhydrite, gypsum and sylvine. Similarly borax crystallizes as a result of the evaporation of water containing boron. Aragonite also belongs to this group of minerals, as does some sulphur, which is produced by the reduction of sulphates by the action of anaerobic bacteria.

Halite (Rock salt)

Halite, usually known simply as salt, has been used as a condiment since antiquity. The simplest means of extraction is by the evaporation of sea-water, which leaves salt as a residue. This method is common in arid countries, and was known to the ancient Greeks. The name derived from *halos*, meaning salt, sea. G Agricola differentiated several varieties of salt, including sea and lake salt, and rock salt. Rock salt occurs in a variety of geological formations, the Zechstein deposits in north western Europe extending from Germany to Britain being of major economic importance, on account of their richness and extent. The order of deposition of the various salts is determined by their solubility in water. The sulphates anhydrite and gypsum are first deposited, followed by the more soluble sodium, potassium, and

Halite with fluid inclusions, cleavage fragment
Leopoldshall near Stassfurt (G.D.R.)
Size: 7cm × 7cm × 2.5cm

magnesium salts. Impurities, especially iron oxides, clay and bitumen, give to the salt a red, grey or brown tinge. When pure, crystallized rock salt is colourless. The dominant crystal form is the cube, and crystals from Allertal in Lower Saxony, and from near Detroit in the United States of America, have an edge length of up to one metre. Octahedra are found more rarely and their development is dependent on the presence of certain ions in the solution.

The perfect cubic cleavage produces cleavage fragments which are often perfect cubes which can be mistaken for naturally-formed crystals. Layers of inclusions are frequently visible in halite crystals. These layers may be formed of solid inclusions, fluids (often with a bubble) or gases. If salt is heated the gases expand considerably and the resulting pressure may cause crazing of the crystal. Another interesting variety is blue rock salt. The blue colour, which varies in intensity and can be a general clouding or distributed in linear patterns, is not caused by impurities, but by the presence of colloidal sodium in the crystal structure. This effect can also be produced artificially by exposure to radiation from radio-active or x-ray sources.

The most important distinguishing characteristics of halite are its ready solubility in water and its salty taste, and the strong yellow colour it imparts to the Bunsen flame, proving the presence of sodium.

Salt is an important raw material for the chemical industry. The electrolysis of molten salt reduces it to its two components, sodium and chlorine, which have many uses both as elements or in the production of other compounds.

Important salt deposits occur widely in Europe, and mostly in rocks of Triassic age. The most extensive British deposits are in Cheshire. Although there is still one working mine, most of the salt is obtained by pumping water into the salt beds, then evaporating the resulting brines.

Gypsum with small cavities
Mansfeld Mulde near Eisleben (G.D.R.)
Size: 10cm × 16cm × 12cm

Gypsum, partially dissolved
Ellrich, Harz (G.D.R.)
Size: 9cm × 11cm

The oldest use of gypsum was recorded by HERODOTUS in approximately 450 A. D. He described how Ethiopians in the Persian army prepared themselves for battle by painting one side of their bodies with gypsum and ruddle. The Greeks were familiar with the cementing properties of crushed gypsum and used it for house building. Yet the reference to the heat generated when the mortar is mixed would suggest that lime was used. The confusion between lime and gypsum continued until the eighteenth century. A F CRONSTEDT solved the problem by the use of acid: chalk effervesces, gypsum does not. Since baryte was also regarded as gypsum, its separate identity as a mineral was established only after barium was discovered by K W SCHEELE.

The most reliable diagnostic properties of gypsum are its low hardness—it is easily scratched with a finger-nail—its perfect cleavage and its poor thermal conductivity—unlike marble, it feels warm to the touch.

Gypsum is mostly white and has a pearly lustre, although the inclusion of clay, sand or iron compounds give it a brown colour, bitumen makes it black, nickel salts will produce green, and sulphur, yellow.

Foliated colourless gypsum (selenite) has always been highly prized. It was known to the ancients as 'mirror-stone,' and in German speaking areas as *Marienglas* (Mary glass). This latter name derived from the use by nuns in convents of wafer-thin sheets of gypsum to decorate and protect the images of their saints from the damaging effects of exposure to air.

Tincalconite on borax
Boron, California (U.S.A.)
Size: 15 cm × 11 cm × 11 cm

Gypsum—twinned crystal
Mazarrón (Spain)
Size: 11 cm × 13 cm × 10 cm

Twinning is common in gypsum, the tabular crystals forming swallow-tail twins with deep re-entrant angles. Such twinned crystals are quite common in the London-clay and can be found at a number of localities along the lower Thames. Crystals are also found in clay at Shotover Hill near Oxford, and at Nantwich in Cheshire. When the Zechstein sea was drying up some 240–225 million years ago, gypsum was among the first minerals to be precipitated as the salt content of the sea water increased owing to evaporation.

There are several varieties of gypsum with different textures. Fibrous gypsum occurs as a fissure filling and consists of parallel fibrous masses with a silky lustre that are orientated perpendicular to the walls. This form of gypsum is sometimes called satin spar and is relatively common in some parts of the south of England.

Compact masses are called alabaster, and were at one time mistaken for marble. Alabaster has been used for carving since ancient times and was presumably popular because of its pure white colour and low hardness which makes it very easy to work. Gypsum can also form through the hydration of anhydrite, a process which results in a considerable increase in volume so that the bedding planes are deformed and contorted.

Gypsum often forms in clays and marls through the weathering of pyrite. The crystals frequently form an aggregate of appreciable size in the form of a rosette. Finally, there are the highly prized 'desert roses.' Individual disc-like plates of gypsum which have crystallized in fine sand can exceed 30 cm across.

The economic importance of gypsum has risen enormously in recent years. It would be hard to imagine the construction industry, for example, without gypsum for the manufacture of roofing sheets and boards. Modern processes in the chemical industry use gypsum in the production of sulphuric acid and cement. Furthermore, it is an important raw material in the manufacture of porcelain and paper.

Gypsum 'desert rose'
Fringe of the Sahara (Morocco)
Size: 13 cm × 12 cm × 8 cm

Borax

Borax was already in use in the Middle Ages to produce glazes and enamels and Venetian merchants transported it from borax lakes in western Tibet, by way of Persia. When freshly extracted, the prismatic crystals range from yellowish-white to grey and are transparent. Exposure to air makes the borax crystals lose their water of crystallization, become cloudy, and in time turn into a white powder.

The most elegant test for boron compounds is the green colour they impart to the Bunsen flame. Further distinguishing features of borax are its low hardness, solubility in water and efflorescence caused by loss of water.

The richest deposits of boron minerals were discovered in Death Valley in California, in 1928. Use of the mineral was extended to the manufacture of permanent, deep-coloured glazes and enamels, and as an additive to certain types of glass. Borax is also used to prevent oxidation of metallic surfaces during soldering, and is used in medicines.

Aragonite

On chemical analysis, aragonite has often been found to contain small amounts of strontium, lead or zinc. For a long time the presence of these elements was thought to be the reason why this form of calcium carbonate crystallized as orthorhombic and not the usual trigonal crystals. Repeated analyses, however, always yielded the same chemical composition as calcite, yet the differences in the physical properties led A G WERNER to recognize the mineral found in the province of Aragon in Spain, as an independent variety. Further investigation confirmed that the crystals from Aragon were indeed an independent mineral. The density and cleavage, as well as the measurement of interfacial angles clearly revealed differences between calcite and aragonite, and so polymorphism, the existence of a chemical substance in two or more crystal forms, was first discovered.

Aragonite seldom forms individual crystals, but mostly occurs in cyclic interpenetration twins in which three crystals combine to give a form with pseudo-hexagonal sym-

172

Aragonite
Iglesias, Sardinia (Italy)
Size: 10cm × 8cm × 8cm

Aragonite, var. flos ferri
Erzberg near Eisenerz (Austria)
Size: 27cm × 7cm × 21cm

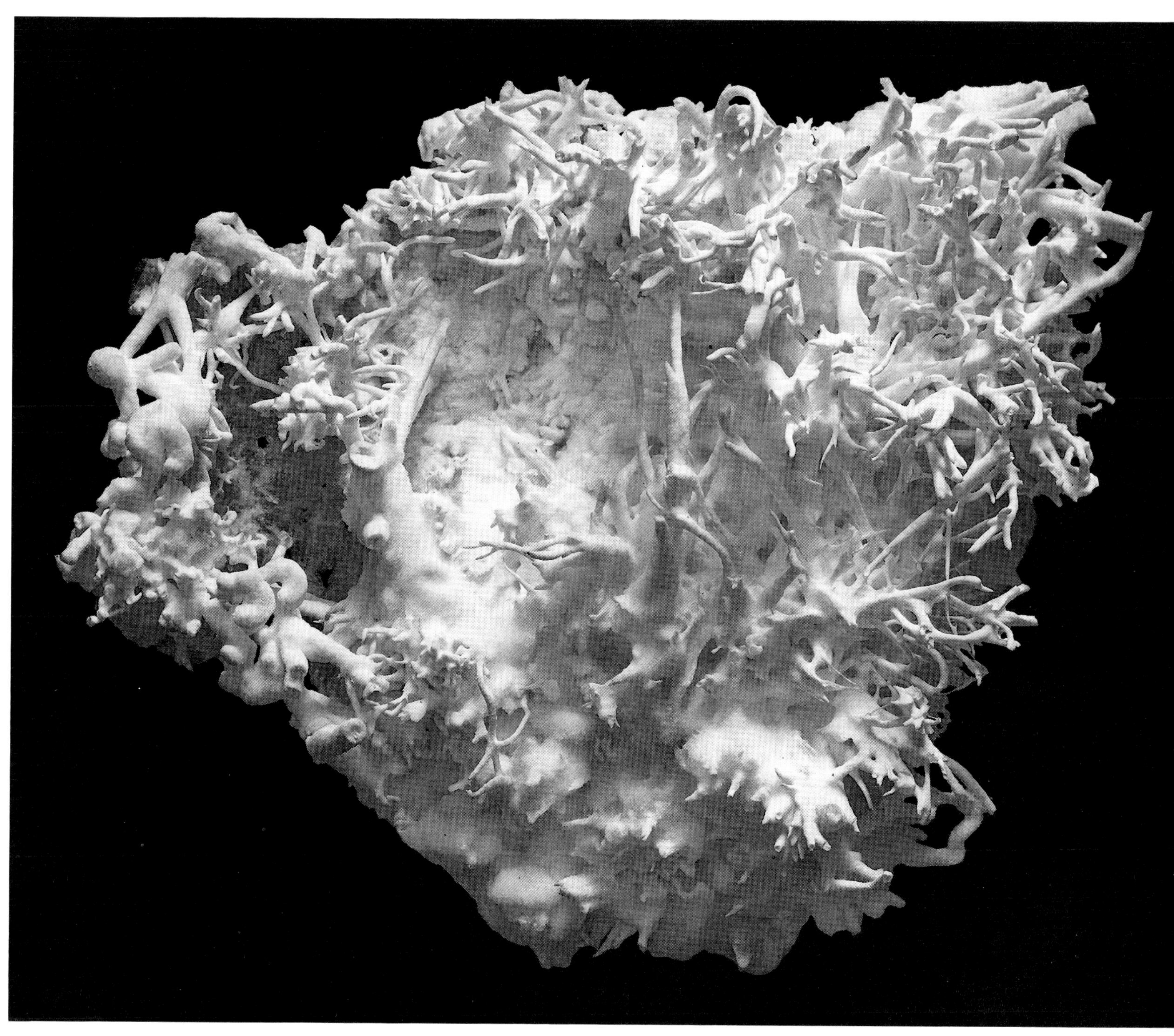

173

metry. The crystals from Bastennes, near
Dax in France, from Molina in Aragon
(Spain), and from Agrigento in Sicily are
typical twins of this type. Groups of acicular
crystals are found at Alston Moor, Frizington
and Cleator Moor in Cumbria. The colour is
extremely variable, but reddish, grey, green
and blue to violet shades predominate.

Individual crystals may be transparent
though usually white to yellowish and they
are elongated along the axis, and have chisel-
like terminations.

Aragonite is also precipitated from hot
springs, predominantly in radiating or fibrous forms. The banded sinters from Karlovy
Vary (Carlsbad) in Bohemia (Czechoslovakia)
are world famous and the pisolitic structures
are known as 'pea stones.' Their low hardness makes them easy to work, and they have
long been used for artistic objects and decorative purposes.

Pseudomorphs of aragonite after gypsum
have a silky, silver iridescent lustre, the original layers of gypsum having been replaced
by aragonite, or partially by calcite, leaving
the form and structures unaltered.

Flos ferri is a particularly beautiful variety
of aragonite. The coralloid, delicate, snow-
white divergent and intertwining branches
form as efflorescences when siderite oxidizes
to limonite and the calcium which is commonly present forms aragonite. Flos ferri
from the Erzberg near Eisenerz in Styria,
Austria, are represented in many collections.

The lustrous mother-of-pearl coating the
inside of many shells as well as the exterior
of pearls is also aragonite. The aragonite
layer is soft and very easily worn away, and
so rapidly loses its lustre. Pearls worn as jewellery must, therefore, be treated with special
care to preserve their beauty.

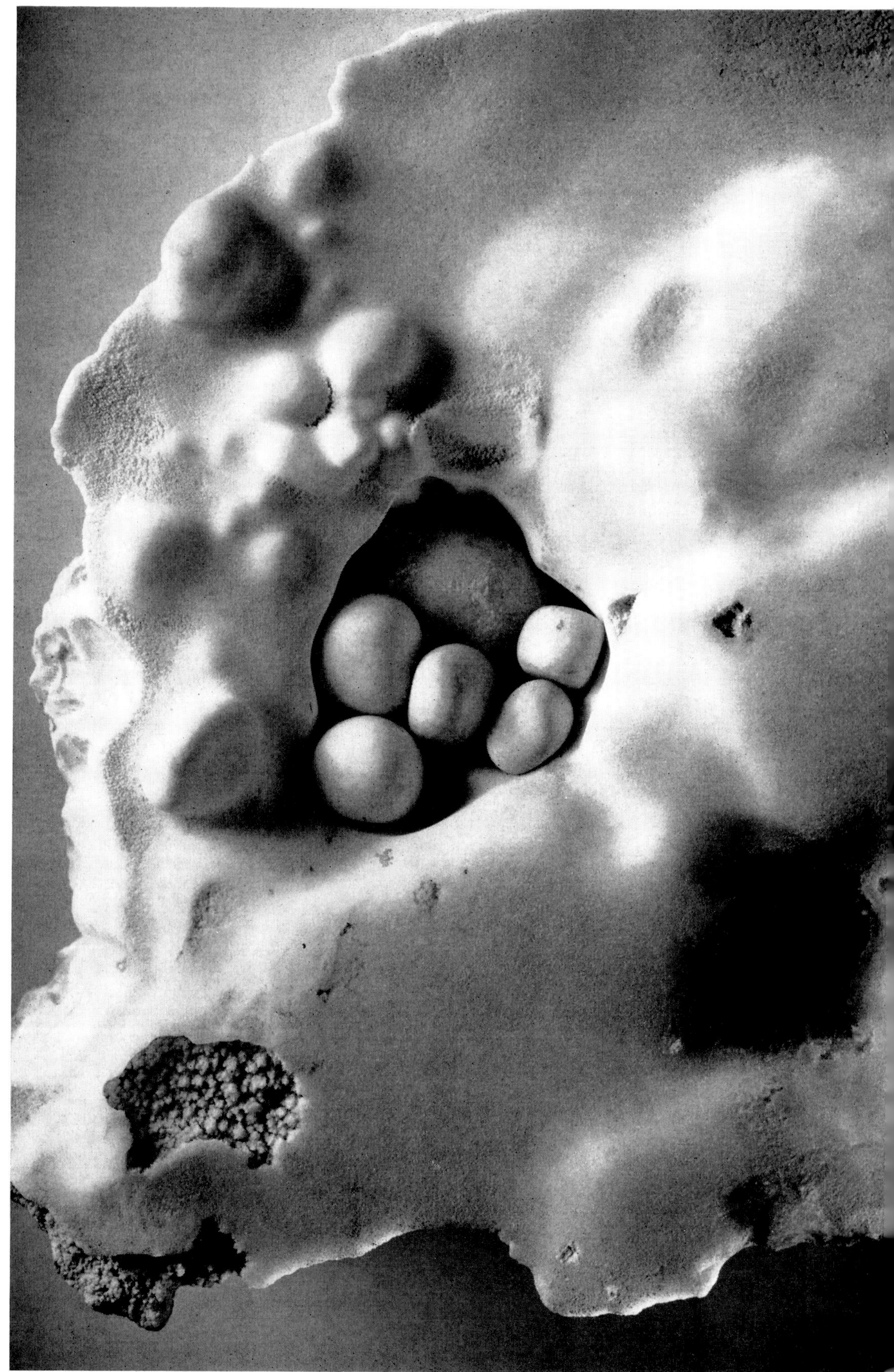

Aragonite so-called calc sinter
Richelsdorf near Bad Hersfeld (Federal
Republic of Germany)
Size: 15 cm × 15 cm × 3 cm

Sulphur with aragonite
Agrigento, Sicily (Italy)
Detail

Sulphur

Sulphur is both a striking and easily recognized mineral which accounts for the fact that it has been known and used since the earliest times. Its most conspicuous feature is its bright yellow colour, which takes on a brownish tinge when it occurs in deposits together with organic compounds such as bitumen and carbon. The crystals, which are frequently of dipyramidal or of tabular habit, have an adamantine lustre. Sulphur is particularly sensitive to sudden, major fluctuations in temperature, which cause fissuring in the crystals and diminish their quality appreciably.

Sulphur has been obtained from Sicily since antiquity. In ancient times operations were limited to extracting sulphur from the fumaroles of Mount Etna, and it was used to kill vermin in the homes of the island's inhabitants.

The prevention of mildew in the vineyards, and the demand for sulphur for gunpowder increased the demand for flowers of sulphur and stimulated production, which reached its maximum at the turn of the century, when there were 730 mines, mainly working to supply the requirements of sulphuric acid production. The introduction of pyrite as a feedstock in the manufacture of sulphuric acid, brought a steady decline in sulphur production, however, so that by 1975 there were only thirteen working mines left.

This decline is making sulphur specimens, so prized by collectors for the large, well-formed crystals, increasingly scarce. The beauty and value of the specimens does not rest merely in their colour, but association with colourless celestine and deep red fluorescent aragonite or calcite increases their value. The closure of the richest mine, near Canicatti, has further increased the value of crystallized sulphur.

In the United States of America, sulphur is obtained on a large scale in Louisiana and Texas. Near Lake Charles in Louisiana, a bed of sulphur occurs which is 100 feet thick, and a similar deposit is found at Freeport in Texas.

*Garnet (surface coated with chlorite);
weight of crystal: 3.75 kg, excellent rhombdodecahedral formation
Falun (Sweden)
Size   14 cm × 14 cm × 14 cm*

Metamorphic Minerals

Metamorphic processes, which occur mainly at considerable depths, result in great changes in the mineralogy and structure of rocks. Depending on the type and degree of the metamorphism, recrystallization may result in alterations in texture, and even melting of the rocks may take place. As a result of these processes, existing minerals may remain unchanged or a mixture of new minerals may form having the same overall chemical composition. Examples of such new minerals are garnet and epidote. Metamorphism also occurs in the vicinity of igneous intrusions which, when they are emplaced, heat the surrounding rock. Minerals produced by this type of metamorphism include brucite, pectolite, some types of amphibole and many others.

Garnet

Garnet is the name applied to a group of minerals, comprising several individual members. The name of these minerals may derive from the colour of some of the more common varieties, which is similar to that of the blossom and fruit of the pomegranate tree. Its flowers are a purplish colour and the fruits are leathery berries filled with red seeds or grains, and the Latin word for 'grain' is *granum*. An alternative explanation of the mineral name could refer to the crystal form of garnet which tends to form richly facetted cubic crystals which have a rounded, near spherical appearance for which the Latin *granum* is appropriate. The commonest crystal form of garnet is the rhombdodecahedron, but it also crystallizes as icositetrahedra or in combinations of the two. The colour of garnet depends on its composition. The predominant red colour results from the presence of iron.

The only colour absent from the broad palette is blue. The range of colour of garnet, as well as its hardness, transparency and lustre, have long made it rank as a gemstone.

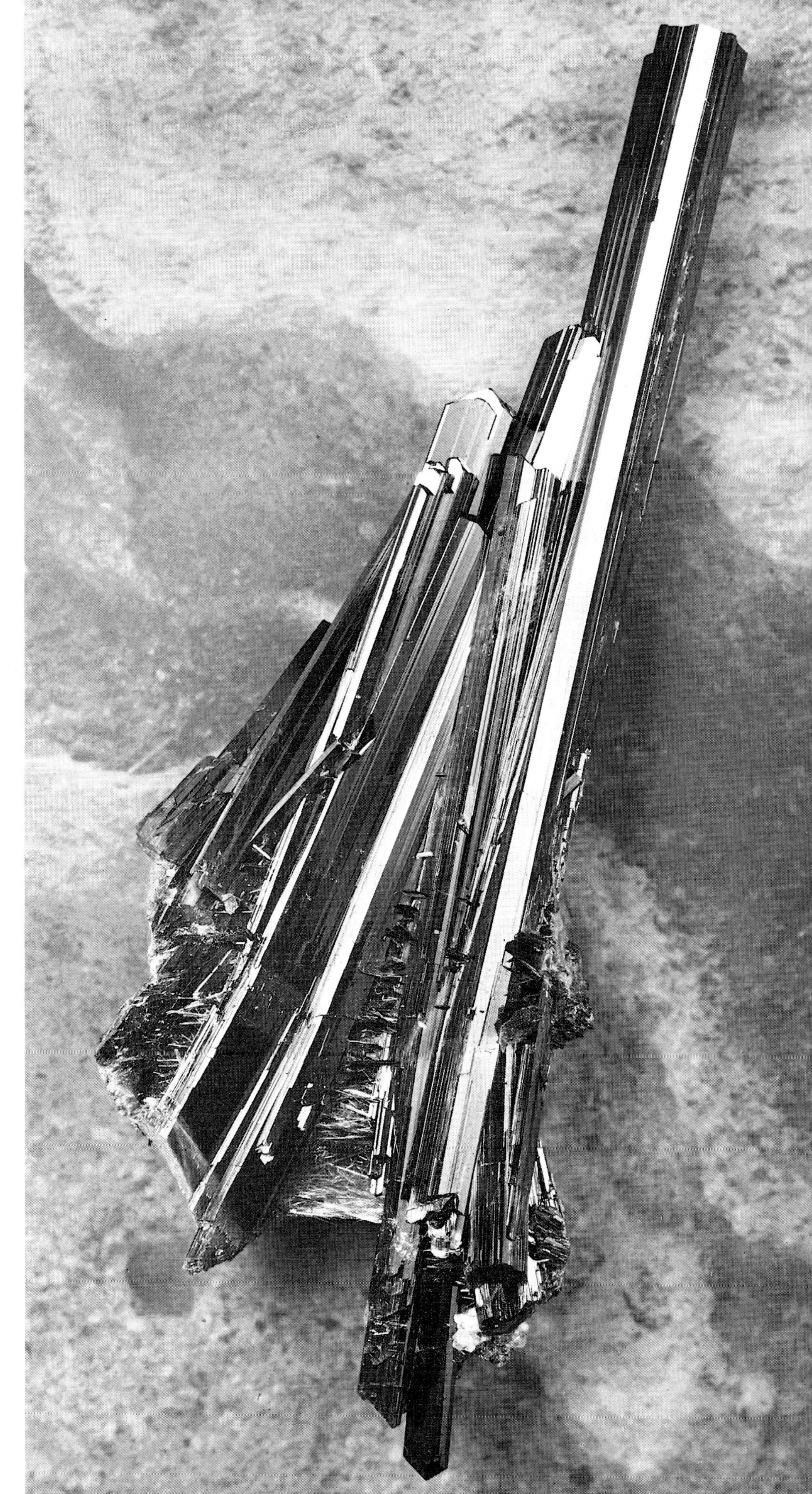

Epidote
Pinzgau, Hohe Tauern, Tyrol (Austria)
Size: 4 cm × 13 cm × 3 cm

Demantoid, the rare green variety of garnet is considered the most valuable.

Natural garnets are largely mixed crystals that fall into two main groups—the pyrope, almandine and spessartine group, and the grossular, uvarovite and andradite group. Within one group there can be continuous variation between the individual members, but only very limited solid solution occurs between the two groups.

Pyrope, also known as Bohemian Garnet or Cape Ruby, is blood red and comes mostly from Bohemia, Czechoslovakia, South Africa and Arizona, United States of America.

Reddish-brown to purplish red almandine is widely distributed in metamorphic rocks (gneisses and mica schists), for example in the Northern and Central Scottish Highlands, in Sweden, Norway, the Alps in Austria, Alaska and on the Island of Madagascar.

Spessartine is usually of orange colour, and occurs in granites and pegmatites, for example in Madagascar, and in Tanzania.

The predominant colours of grossular are grey, reddish brown, pale green and raspberry red. It occurs in Brazil, Tanzania, Canada and the Transvaal.

Emerald green uvarovite occurs in chromium deposits in the Urals, Soviet Union; the Transvaal; South Africa and in Finland and the green colour is caused by the high chromium content.

Black garnets which are rich in titanium and called melanite, occur in alkaline igneous rocks in the Kola peninsula, Soviet Union; Arkansas, United States of America and Rusinga Island, Kenya. It is also found in the Fen area of Norway and on the island of Alnö, Sweden.

Green andradite or demantoid is found in the Urals, Soviet Union, in the gold deposits of Nizhni Tagil and in serpentinite near Sysert.

Chrysotile asbestos with serpentine
Montville, New York (U.S.A.)
Size: 25 cm × 13 cm × 5 cm

Chrysotile Asbestos
(Serpentine Asbestos)

Chrysotile asbestos—or simply, asbestos—is a
fibrous variety of serpentine, which is worked
extensively for industrial use. The asbestos
fibres are aligned across the veins of serpentine. The pale to golden yellow colour of the
fibres, whose silky lustre makes them gleam
in the sunlight, gave the mineral its name.
Translated from the Greek, chrysotile asbestos means non-burning, golden fibres
(*chrysos*—golden, *tilos*—fibre, *asbestos*—in-
fusible). The colour can also be greenish
yellow, or brownish, and very occasionally
the fibres are white.

The value of the asbestos depends on the
length of the elastic, but strong fibres: these
are usually between 1 and 2 cm long, and
only rarely exceed 10 cm.

Very long fibres can also occur along the
shear planes of serpentinite, but these have
no commercial value.

Asbestos has been known since antiquity.
The wick in the golden lantern of Minerva
was asbestos. PLINY describes 'asbestos' whose
value equals that of the finest pearls, that
'grows' in deserts and sun-baked regions of
India, where no rain falls, and which is hard
to weave because it is so short.

The oldest known asbestos deposit is on
the Island of Cyprus. The world's commercial
requirements of asbestos are met by deposits
in the neighbourhood of Turin in Italy,
Thetford in Canada, Barberton in the Transvaal, South Africa, Shabani in Zimbabwe-
Rhodesia, Bashenovo in the Urals, Soviet
Union, and from east and west Sayan in the
Soviet Union.

Amphibole Asbestos (Amianthus)

Amphibole asbestos, sometimes called amianthus, is distinct from chrysotile asbestos
though it shares with it its characteristic in-
fusibility. Amianthus is a variety having fine
and flexible fibres with a silky lustre. It forms
delicate coatings in Alpine joint cavities, and

*Amphibole asbestos with rock crystal and
chlorite
Teufelstal near Andermatt, St Gotthard
(Switzerland)
Size: 18 cm × 9 cm × 5.5 cm*

clusters of fine, hair-like fibres embedded among rock crystal or smoky quartz, which are also encrusted with a deep greenish-black dusting of chlorite.

Over the centuries the form of amianthus gave rise to many popular names in mineral-ogical literature. 'Mountain wool' refers to the loose matting of very flexible fibres. 'Mountain wood,' by contrast, has a firmer, stiffer structure. The compact, tangled inter-twined form is known as 'mountain cork,' and the whitish-grey, densely matted layers, are called 'mountain leather.' It is possible to make paper from asbestos fibres and several copies of a treatise on asbestos were printed on such paper, in demonstration of its fire-resis-tant properties and durability. The name amianthus was used by the ancient Greeks—

amianthos means pure, unstained; a broad reference to its distinctive characteristic of resistance to fire, for it is possible to throw it into fire to clean it.

Brucite

Brucite usually forms scaly, curved crystals with a silky to waxy lustre which are soft and flexible. The colours vary from white to grey, and bluish and greenish shades, depending on the presence of impurities.

Brucite from Hoboken in New Jersey, United States of America, was first described in 1814 by the American mineralogist, A BRUCE. In 1824, F S BEUDANT named the mineral brucite, in honour of its discoverer.

One of the richest deposits, largely associated with serpentine, is at Asbestos in Quebec, Canada, where the brucite fibres are up to 50 cm long. Brucite occurs in veins in serpentinite, as a result of the hydrothermal alteration of magnesium-rich rocks such as dunite. It is also formed by the metamorphism of dolomitic limestones and such brucite marbles are to be found in the Assynt area of northwest Scotland, and on the Island of Skye. Brucite is used in the purification of crude oil distillates, the organic sulphur compounds (mercaptane) being retained by the brucite.

Epidote (Pistacite)

Elongated, prismatic crystals of epidote were first described by a French mineralogist, R J HAÜY, in 1801. The name derives from the Greek word, *epidosis* meaning 'increase.' It had been regarded formerly as a variety of schorl (tourmaline), and had repeatedly been mistaken for actinolite, an amphibole, but HAÜY pointed out the differences in crystal form between epidote and actinolite. In epidote, the cross section has the form of a paral-

Asbestos (chrysotile asbestos)
Amianthus mine, east Transvaal (South Africa)
Size: 12 cm × 13 cm × 8 cm

Brucite
Texas, Lancaster County, Pennsylvania (U.S.A.)
Size: 11 cm × 12 cm × 3 cm

lelogram (with sides of different lengths),
whereas in amphiboles it is usually a rhombus. The 'increase' alluded to in the name
epidote is that of the elongation of two sides
of a rhombus to form a parallelogram.

The world-wide distribution of epidote
meant that many varietal names were used
to describe colour variations, or to honour
their discoverers or particular localities.
A G WERNER chose the name 'pistacite' (from
the Mediterranean evergreen, *pistachio*) for
green epidote; varieties from Arendal in
Norway and Akhmatov in the Urals, Soviet
Union, were named arendalite and achmatite,
while the red manganese-bearing epidote
from Glencoe in Scotland was called with-
amite in honour of its discoverer, H WITHAM.

Epidote from Knappenwand, a part of the
lower Sulzbachtal in the Pinzgau, Austria,
has since its discovery in 1866, effectively
served as the standard of quality, being unsurpassed in size, lustre and colour.

Crystals up to 45 cm long, and crystals
3–4 cm thick are recovered from many different localities. It is not just the slender, lath-
shaped crystals from Knappenwand with
their blackish-green colour which are translucent in strong light that are so much admired, but also their arrangement on the
specimen. The epidote crystals stand out from
a felted mass of grey-green, hair-like asbestos needles, in association with white
adularia, colourless apatite, and superficially
corroded calcite. Other well-known occurrences are on the west coast of Canada, in the
Zillertal, Tyrol, Austria (acicular wine-
yellow crystals) and brown, radiating clusters
on the granite in Strzegom in Poland.

Pectolite with calcite
Bergen Hill, New Jersey (U.S.A.)
Size: 7.5 cm × 10 cm × 5 cm

Wavellite on slate
Ronneburg, Thuringia (G.D.R.)
Size: 11 cm × 4.5 cm × 3 cm

Pyrophyllite
Tres Cerritos, California (U.S.A.)
Size: 8 cm × 4 cm × 3 cm

Pectolite

This mineral was first described in 1828 by F von Kobell from the northern slopes of Monte Baldo, on the eastern shore of Lake Garda in Italy. Milky-white radiating spheres occurred in the veins of an earthy basalt tuff, showing when broken, a silky to vitreous lustre. The structure of the hitherto unknown mineral prompted its discoverer to name it pectolite, from the Greek words, *pektos* (joined, fused together) and *lithos* (stone). F A Breithaupt discovered a further, curious characteristic of the mineral, namely that broken surfaces phosphoresce in the dark. This property prompted him to suggest the name, 'photolite' (from the Greek *phos*—light, and *lithos*—stone), although this was not adopted.

The radiating form of pectolite is particularly striking in specimens from Bergen Hill in New Jersey, United States of America. Near translucent radiating fibres—up to 10 cm in length—have a bluish white colour in sunlight, which contrasts with the dark diorite matrix. This contrast makes the specimens particularly attractive and they have consequently caught the attention of collectors.

There are further occurrences of this none too plentiful mineral in the Federal Republic of Germany, at Niederkirchen in the Rheinland-Pfalz and Haslach in the Black Forest, in Želechy near Harrachov in Czechoslovakia and in Prince Charles' cave on the Island of Skye in Scotland. There are also occurrences from Weardale, County Durham and Rathie near Edinburgh. Pectolite from Alaska, near Point Barrow, was used by the Eskimos for making into tools: its compact, tough structure making it eminently suitable for this purpose, proof of which lies in the durability of the articles.

Magnetite (Magnetic Iron Ore)

The wide distribution of magnetite makes it one of the most important ores of iron, along with hematite and limonite. It has been known since antiquity, and its name is supposed to go back to the fable of Magnes, the

Magnetite with dolomite and calcite
Traversella, Piemont (Italy)
Size: 8 cm × 10 cm × 4 cm

shepherd. Legend has it that the magnetic effect of the mineral made the nails in his shoes and the metal ferrule on his crook cling to the ground of Mount Ida, near Troy in Greece. It is more probable that the name is derived from Magnesia, a locality in Thessaly in Greece. The latter explanation indicates that the term, magnesian stone, was ambiguous, and referred to several minerals.

The ancient Greeks divided minerals that looked alike into a male and a female 'magnes,' of which the former could attract iron.

Seafarers were using magnetite in the twelfth century to make simple compasses, and consequently it was also known in German as 'Segelstein' (*Segel* meaning 'sail'). The 'female magnes' was used mainly to bleach molten glass, and to produce brown glazes in pottery. Manganese ores resembling magnetite, mainly pyrolusite, were also used for the same purpose.

Furthermore, a white, non-magnetic, female 'magnes' was used in medical remedies. It is impossible to say for certain which mineral was meant by the term 'magnesia alba,' but it is thought to have been alabaster or talc.

Apart from its deflection of the compass needle, magnetite is easily recognized by its iron-black colour, which takes on a blue-black sheen in the light, and by its metallic lustre. The dominant crystal form is the octahedron; rhombdodecahedra are rarer, and recognizable by their striations.

Display specimens with fine, large magnetite crystals come from the Traversella deposits in Italy, from Kovdor on the Kola peninsula and the Vysokaya mountains near Nizhni Tagil in Dana and Blagodat near Kushva in the Urals in the Soviet Union. Further well-known localities are the Zillertal in the Tyrol, Austria, Binnatal in Switzerland, and Magnet Cove in Arkansas in the United States of America. The strongest natural magnets of magnetite come from Siberia, the Harz Mountains, and from the Island of Elba.

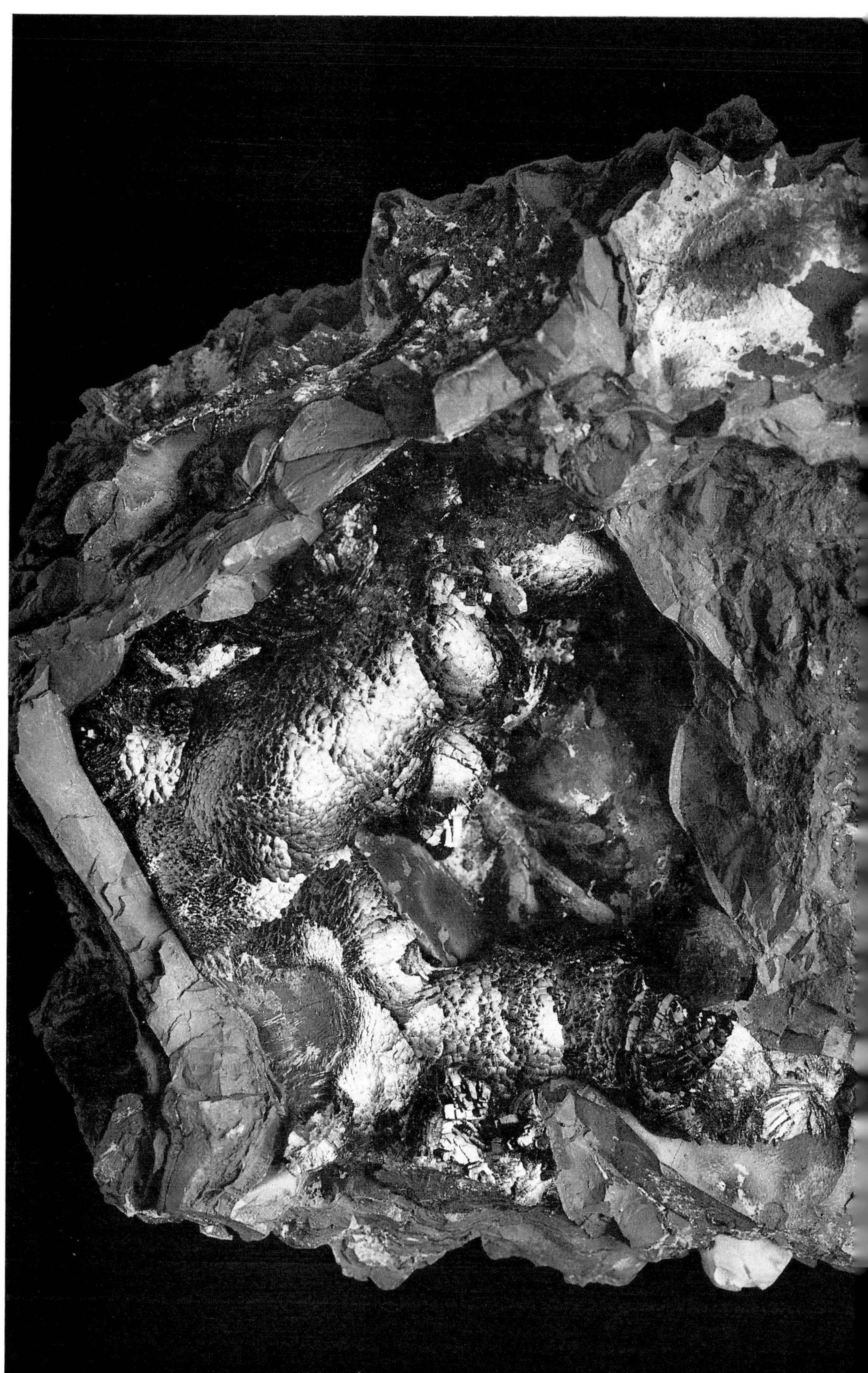

Groutite on limonite
Bülten, Lower Saxony (Federal Republic of Germany)
Size: 9 cm × 8 cm × 2 cm

Newly Discovered Minerals

New minerals are being discovered all the time. Generally they are encountered accidentally, perhaps by people working on Museum collections, or during evaluation work on mineral deposits. The total recognized by the International Mineralogical Association (IMA) after exhaustive investigation amounts to between thirty and forty every year. Naturally the commoner minerals are those which are distinctive by virtue of their colour or some striking physical feature, which were described long ago, so that new minerals are now rather more difficult to discover and many are found by the application of newer techniques such as the electron microprobe. Many new minerals only occur as very tiny masses or crystals which can only be seen with the microscope. The electron microprobe is a machine with which it is possible to chemically analyze single mineral grains as they are seen down a microscope. In the past it was necessary to separate a pure mineral fraction before a chemical composition could be determined: this is no longer the case. However, many of these newly described minerals are only to be seen with a lens or microscope, so they are of little interest to most collectors.

Investigation of old, abandoned mines has brought to light a number of minerals which were previously unknown. The detailed investigation of specimens in mineralogical collections is another way in which new minerals are discovered. Space research has opened up entirely new vistas in this field, and a number of new minerals have been found in lunar rocks.

The criteria for the recognition of a new mineral are laid down by the New Minerals and Mineral Names Commission of the IMA. These include a chemical analysis and formula, physical characteristics—hardness, lustre, specific gravity, cleavage, fracture and

Charoite with tinaksite
Mount Murun near Kadar in the valley of
the river Charo, Siberia (U.S.S.R.)
Detail

188

refractive index, x-ray analysis with a calculation of the cell dimensions and determination of the crystal system, together with details of any associated minerals (paragenesis). Suggestions for the name of the new mineral continue to relate often to the discoverer or some other person, the locality or to the composition of the mineral. The Commission tries to ensure that new names are not synonyms, are pronounceable and are generally suitable.

Charoite

Charoite is a mineral that was discovered in 1960 in the alkaline complex of Murun in Siberia and was described by Y G Rogov, a Soviet geologist. It is a complex alkali silicate named after the Chari river.

The mineral is violet in colour, and forms clusters of radiating fibres with a silky lustre which is enhanced when specimens are polished. Charoite resembles a metamorphic rock in having a foliated structure. The foliated structure of charoite, its association with yellowish-brown tinaksite (which differs only slightly from it in composition) and the fibrous intergrowth of the two minerals are responsible for the toughness of specimens. Aegirine and canasite also occur in the assemblage. Grinding and polishing greatly enhances the decorative appearance of charoite.

Externally charoite resembles certain amphiboles, the fine, fibrous appearance being reminiscent of asbestos, whilst its greasy feel is comparable with that of serpentine.

Scholzite with collinsite on limonite
Reaphook Hill near Blinman, Flinders
Range (Australia)
Size: 12 cm × 6 cm × 6 cm

Scholzite

This new mineral was first found in 1949 by A SCHOLZ, and was named in his honour by H STRUNZ. It forms small, thick, tabular or bladed, white to colourless crystals associated with muscovite and is further intergrown with quartz, feldspar, apatite and dark, massive sphalerite. Scholzite, a calcium-zinc-phosphate is formed as a secondary mineral in pegmatites by reciprocal reaction between solutions and apatite and sphalerite. Finely crystalline coatings of the new mineral occur mostly in cavities in corroded apatite.

Scholzite was referred to by MÜLLBAUER in 1925 as a white phosphate mineral from the same locality.

Considerably larger crystals of scholzite, the best specimens known to date, have been found in recent years at Blinman in South Australia, as clusters of acicular crystals 2–3 cm long. Their brilliant white colour makes them stand out vividly against the brown limonite matrix. Here and there scholzite crystals are coloured deep black by manganese compounds.

Further examples of newly discovered minerals are tuscanite and franzinite, found in Tuscany, in Italy, and keyite and leiteite from the world-famous Tsumeb mine in Namibia.

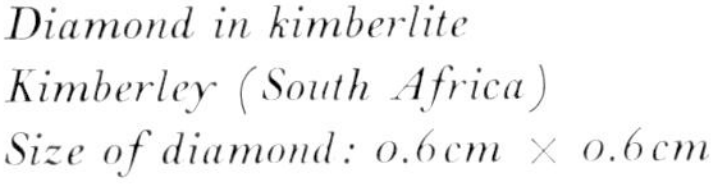

Diamond in kimberlite
Kimberley (South Africa)
Size of diamond: 0.6 cm × 0.6 cm

Precious opal
Queensland (Australia)
Size: 5 cm × 5 cm

Precious and Semi-precious Stones

The aesthetic appeal of many works of art rests to a large extent on the precious and semi-precious stones they contain. Of some 2,500 known minerals and varieties, only about seventy are considered exceptional. This select group has remained essentially the same for thousands of years, but the search for new occurrences continues unceasingly, as does the further exploration of known deposits. The particular properties that single out a 'precious stone' are its hardness, colour, transparency, lustre and special inclusions (as in amber). Those minerals which are less valuable or not transparent, such as agate and jasper, are called 'semi-precious stones.' Common to both precious and semi-precious stones are their beauty, durability and rarity.

The beauty of precious stones is exemplified by the clarity and brilliance of diamond, the play of colours shown by opal, the patterns of agate, the strong colours of lapis lazuli and malachite, and the lustre of topaz.

Precious and semi-precious stones retain their beauty because of their resistance to wear and use. It is the hardness of minerals that preserves the colour, form and lustre of cut stones for so long. Hardness is a consistant physical property which is frequently used to identify a mineral. There is a distinction between the scratch hardness associated with the name of F Mohs and the grinding hardness of A Rosiwal. Scratch hardness is the resistance offered by a mineral when scratched with a sharp object. The hardness scale comprises ten different minerals. A mineral in the scale will scratch any mineral of lower hardness, and will itself be scratched by any mineral higher in the scale. The choice of minerals is somewhat arbitrary in that diamond, with a hardness of 10, is not eight

Topaz with quartz and nacrite
Schneckenstein, Vogtland (G.D.R.)
Size: 11 cm × 9 cm × 9 cm;
size of the crystal: 1.2 cm × 1.3 cm

Ruby on marble
Mogok (Burma)
Size: 3.8 cm × 3.8 cm

times as hard as gypsum with a hardness of
2. The ROSIWAL grinding hardness indicates
that diamond is 140,000 times harder than
gypsum.

Mohs' Hardness Scale (scratch hardness)

Hard-ness	Mineral	Gemstones
1	talc, graphite	
2	gypsum, halite	amber
3	calcite	malachite
4	fluorite	rhodochrosite
5	apatite	turquoise, soda-lite, lapis lazuli
6	orthoclase	moonstone, opal, tiger's-eye, jadeite
7	quartz	amethyst, garnet, agate, tour-maline
8	topaz	spinel, aqua-marine, emerald, ruby, sapphire
9	corundum	
10	diamond	

Some gemstones occur at only one locality.
The conditions necessary for their formation
obtained at this one place only and as a result
it has become a collector's item on account of
its rarity. An example is the blue zoisite
known as tanzanite.

Only a few of the many ways in which
gemstones are used can be mentioned here.
Secular and religious leaders have sought to
display their power and wealth by building
up treasure houses. Their riches comprised,
besides the precious metals gold and silver,
crown jewels, magnificent jewellery, rare
coins and richly-decorated and trimmed robes
of office.

The armoury in the Kremlin in Moscow
contains treasures that can hardly be surpassed dating from the time of the Czars. The
outstanding items in this collection are the
'Shah' and 'Orlov' diamonds, and the glitter-
ing crowns and diadems. The wealth of the

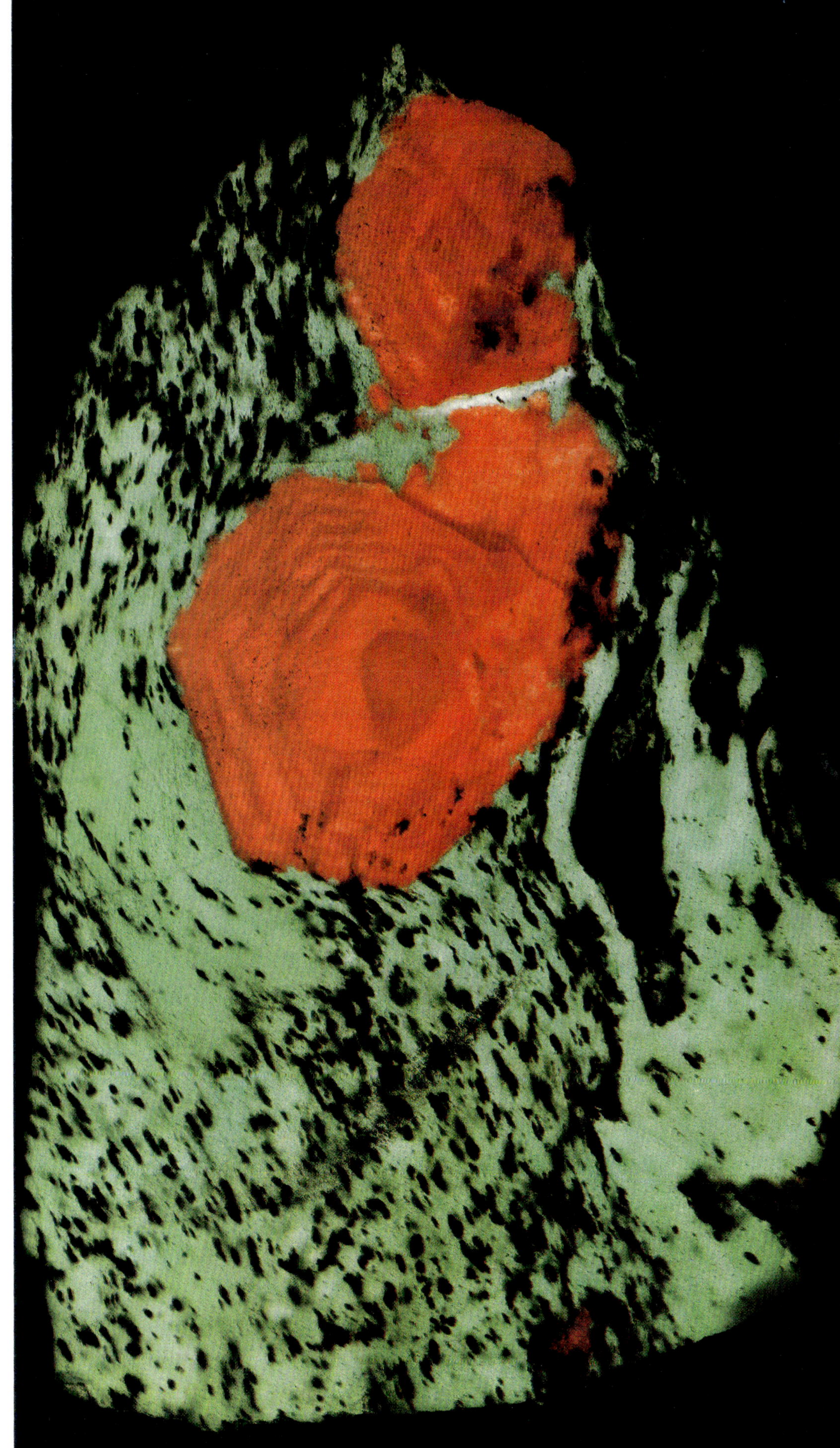

Ruby in zoisite
Matabata Hills, Kilimanjaro (Tanzania)
Size: 15 cm × 21 cm × 0.2 cm;
size of the included ruby crystal:
6.5 cm × 7 cm

195

Court of Saxony was, however, able to compete with this splendour; in about 1560 the *Grünes Gewölbe* (Green Vault) was established in Dresden as a treasure chamber and was continuously expanded thereafter. In the years from 1694 to 1733 during the reign of Augustus the Strong, many pieces were added to the collection of art treasures. Most of these were the work of the famous jeweller, J M DINGLINGER. The *Grünes Gewölbe* houses over 3,000 objects, of which only a portion are on show. The sumptuous design and constant expansion of this art collection were made possible only because of the lucrative mines in the Erzgebirge in Saxony with their rich deposits of silver and gemstones, and by the ruthless exploitation of the miners and smelters. By the middle of the eighteenth century, the collection had reached its peak. One mineralogical treasure known throughout the world is the Dresden green diamond.

The treasuries of the ruling houses were enormously wealthy with their collections of gems, precious metals and curiosities. In times of need these valuables were converted into actual coins. Gemstones have the advantage of concentrating high value within a small space, and they are a stable investment commodity, more so than many other items. Precious and semi-precious stones, particularly if they have any notable historical associations, have held their value well over the past decades.

The treasuries of rich men are not the only places where precious and semi-precious stones are to be found, there are collections also in museums and in universities and colleges. In these collections it is not unusual for single crystals of gem minerals, either alone or on matrix, to be displayed alongside the cut stone, so giving further information, especially about the relationship between the mineralogical properties and the way the stone can be cut. The Smithsonian Institution in Washington, the British Museum (Natural History) and the Geological Museum in London, the Naturhistorisches Museum in Vienna, and the École des Mines in Paris all possess and display excellent collections of gemstones.

Spodumene, var. kunzite
Hambacuri, Minas Gerais (Brazil)
Size: 8cm × 20cm × 3cm

Diamond

Gold has been called 'king of metals' and diamond deserves a similar title in the realm of precious stones. It combines exquisite beauty and purity, legendary lustre and great durability and so is of the highest value. All these properties are closely related to its physicochemical structure, giving it an exceptional position among minerals. The singular hardness of diamond was known in antiquity. PLINY wrote of it: "Diamonds can be recognized on the anvil, because they return a blow, so that the hammer cleaves in two, and even the anvil itself may split. The hardness, thus, is immense, moreover this variety defeats fire, for it cannot be warmed." This is how it came to be known as the 'insuperable force'—in Greek, *adamas*—which, according to legend, could only be broken by fresh, warm goat's blood.

Some of the finest and most famous diamonds came from India (eastern India). The ancient Indian Sanskrit literature, especially the sixth century book of Brahat Samhita, contains references to diamonds and where they are found. Diamond is found in stream sediments by reason of its hardness and resistance to abrasions. The crystal edges become blunted by transport and the crystal faces become matt and rough, and such diamonds are difficult to distinguish from quartz.

The most important localities in India lay along the Pennar, Kistna, Mahanadi and Brahmani rivers in the east of the Deccan highlands. Until the tenth century, all diamonds remained in India, either in the treasuries of the princes, or in the temples, as ornaments for the images of the gods. Indian gemstones became known through the accounts of the thirteenth century Italian explorer, Marco Polo, and those of the Portuguese, Garcias Horto, dating from 1565.

Euclase on mica
Lukangasi, Morogoro District (Tanzania)
Size: 7cm × 4.3cm

Quartz, var. wood opal
Clover Creek, Lincoln County, Idaho
(U.S.A.)
Size: 13cm × 10cm × 4cm

There is a fairy-tale quality to some of the stories surrounding the largest and finest diamonds. One of the most famous diamonds is the Koh-i-Nor ('mountain of light'), which is said to have had an original weight of over 700 carats (1 carat = 0.2 gram), but it may have been much less. This diamond is supposed to have been found in 56 B.C. In the fourteenth century it was owned by the family of the Rajah of Malva, and thereafter, it changed hands several times. In 1739, when Delhi was taken by Nadir Shah of Persia, the diamond was owned by the Great Mogul of India, Mahommed Shah. When it came to sorting the booty, the diamond was missing, and it was only by the intrigue of a woman of the harem that it fell into the hands of the Persian ruler. In the course of a banquet, Nadir Shah offered his Indian guest an exchange of turbans—an oriental custom—and the Koh-i-Nor passed to the Persian. Later, however, it returned to India, but in 1850 the East India Company presented it as a gift to Queen Victoria, at which time it weighed about 190 carats. In 1852 the Koh-i-Nor was re-cut in Amsterdam in thirty-eight days and reduced to 108 carats; it was worn by Queen Victoria as a brooch before being set in one of the British State crowns.

J B Tavernier, the French merchant and jeweller, whose descriptions of his travels in India made him world famous, brought back a great rarity for Louis XIV, a blue diamond of 67.5 carats. In 1792 it was stolen and has never been found. A stone of a similar type weighing 44.5 carats turned up in 1830, but as a result of re-cutting it was impossible to prove beyond doubt that this was part of the stolen diamond. It has gone down in history as the 'blue Tavernier' or 'Hope' diamond, after a subsequent purchaser, H P Hope, the London banker. In 1958 this blue diamond

197

was presented to the Smithsonian Institution in Washington by H WINSTON of New York, who bought it in 1949

The 'Orlov' diamond is also of Indian origin, and was first cut by the Hindus. It was found about the beginning of the seventeenth century, and originally weighed some 400 carats. This diamond is also known as the 'Amsterdam diamond', and is considered the best of all the Indian diamonds. There is a legend that it originally formed the eye in the idol of the goddess Sri Ranga in the temple of Mysore. A French soldier, disguised as a monk, stole it from the temple. Count G G ORLOV bought this diamond in Amsterdam for the Czarina Catherine II in 1774. Since then, as a rose-cut stone now weighing 194 carats, it has been adorning the top of the Russian Imperial sceptre and is now in the Diamond Treasury of the Soviet Union.

The list of Indian diamonds and their history could easily be extended. In the Orient their names are linked with rulers and their estates, in Europe they recall the sovereigns of dukedoms and principalities. By the beginning of the eighteenth century, when the Indian deposits had been exhausted, the search for diamonds moved to Brazil. The prejudice against gemstones from the New World meant that the market accepted them only reluctantly. The best-known gems are the 261-carat 'Star of the South' from Minas Gerais, and the 'Presidente Vargas,' weighing 726 carats, from the São Antonio river, which after cutting yielded 29 individual stones.

Diamonds from South Africa revived the diamond market to an unprecedented extent. The finding of the 83.5-carat 'Star of Africa' in 1869 started off the feverish hunt for diamonds along the Orange River in South Africa. The largest diamond to date is the 'Cullinan' stone, as large as a fist and weighing 3,106 carats, which was found in 1905. The rough stone was cleaved into nine larger and ninety-six smaller pieces for cutting.

In recent years the diamond fields between Lena and Yenisei in Yakutsk, Soviet

Precious opal
Bulla Creek, Queensland (Australia)
Size: 3.5 cm × 5 cm × 1.5 cm

Union, have been gaining considerably in importance. Although they produce mainly industrial diamonds there have been finds of gem quality, among which the 106-carat 'Maria,' found in 1967, and the 'Stalingrad' of 166 carats, are outstanding examples.

The commonest crystal forms in diamonds are the octahedron, the cube and the rhomb-dodecahedron. Many diamonds are etched as a result of reaction with magma, and the edges and corners of the crystals are frequently rounded. The presence of trace elements such as nitrogen, manganese and aluminium in the structure affects the colour, crystal form, cleavage and conductivity. Diamond is the only gemstone to consist solely of one element, carbon, and is recognizable by its extreme hardness which derives from strong covalent bonding forces between the constituent carbon atoms. Its high refractive index and dispersion and lustre further add to its value as a gemstone. The so-called brilliant cut exploits the high dispersion to the maximum effect.

Ruby

PLINY's assertion that in gemstones all majesty is compressed into a tiny space, and that a single gem suffices to make us recognize the miracle of the creation, applies without reservation to the blood-red ruby, 'lord of gems.' The Indians long regarded the ruby in this way. The glorious red colour also gave the gem its name, the origin of which lies in the Latin word, *ruber* (red). The ruby owes its high place in the realm of gemstones, second only to the diamond, to its properties of fascinating colour, extreme hardness and excellent refraction. A member of the corundum group, its chemical composition is aluminium oxide, and the colour, reminiscent of the red glow of a furnace, is produced by the trace amounts (0.1 %) of chromium oxide. Traces of iron change the colour of ruby to a brownish hue, and vanadium turns it blue. Blue corundum of gem quality is called sapphire.

Tourmaline, var. rubellite
Pala, San Diego County, California
(U.S.A.)
Size: 6cm × 13cm × 5.5cm

199

The presence of inclusions is a feature in
the recognition of ruby. Orientated inclusions
of rutile needles give rise, when the stone is
illuminated by a strong point light source, to
a star effect on the polished surface (star sapphire), or the inclusion of small drops of fluid
can form a delicate cloudy bloom.

The richest deposits of ruby are the placer
or gem gravels deposits, for ruby weathers
out of parent rocks, such as marble, dolomite
or crystalline schist to be deposited in stream
sediments. Its high density (3.9–4.1) is an
important factor in causing its deposition
after being transported for only a short dis-
tance. It also serves to separate the ruby
from the clay and gravel matrix.

Rubies have been mined for thousands of
years from the placer deposits of Mandalay
and Mogok in Burma, from Battambang in
Cambodia, Chanthaburi in Thailand and near
Ratnapura on the Island of Sri Lanka. The
richest deposit at Mogok was supposedly, so
the story goes, discovered by a band of con-
victs in the fifteenth century, who were serv-
ing their sentence in this river valley. Recently the ruby finds in Tanzania have been
attracting considerable attention. These
gems are mined in the valley of the River
Umba and in the Matabatu Mountains. Zoned
hexagonal crystals, from Longido, associated with dark green zoisite, are used for
many different types of jewellery. The fine
structure of the ruby can be clearly seen in
direct sunlight in thin, millimetre thick
plates. The rubies embedded in marble from
Prilep in Jugoslavia and those from Campo-
lungo in the Ticino in Switzerland are of a
very inferior quality.

The largest rubies found to date are the
138-carat Rosser Reeves ruby, the 100-carat
Delong ruby and the 42-carat Peace ruby.
Rubies weighing between five and ten carats
or over are exceptional.

Lizard in amber
Baltic coast, Kaliningrad region (U.S.S.R.)
Size: 3.5 cm × 5 cm

Labradorite (cut and polished)
Pauls Island, Labrador (Canada)
Size: 13 cm × 12 cm × 3 cm

The many shades of colour and the hardness of topaz have earned it a place as a gemstone in crown jewels and treasure-houses all over the world. The colourless, 1.680-carat 'Braganza diamond' in the Portuguese crown is a topaz. Although there are many different properties to be taken into account in distinguishing between 'topaz and diamond,' the two are almost identical in one particular aspect, namely that of specific gravity.

Frederick Augustus II of Saxony, son of Augustus the Strong, valued the topaz very highly. He purchased the Schneckenstein in the Vogtland, location of the beautiful, wine-yellow topaz, and from 1737 to 1751 had the brilliant gems mined there. The finest and largest topazes were used for his sumptuous jewellery and his topaz collection can still be seen in the *Grünes Gewölbe* in Dresden. Furthermore, 485 topazes from this famous deposit in Saxony were used in the British crown. The yield of stones suitable for cutting is in fact small, since inclusions of ilmenorutile, fissures and gas bubbles frequently occur, thereby reducing the quality considerably.

The derivation of the name, 'topaz,' is somewhat uncertain. There are references in ancient literature to the island of Topazus in the Red Sea. It was an island shrouded in mists, which seafarers had difficulty in finding, and in their tongue they called it 'Topazin,' meaning 'to search.' There may well be some confusion about the gemstone found on this Red Sea island. PLINY describes a green mineral that occurs in lumps and can be facetted with a file. This would be true of green fluorite, although the likelihood is that he was referring to olivine, which occurs on St John's Island in the Red Sea. The issue is further complicated by the fact that of old, topaz was always known to be a yellowish-brown colour. Thus it could be that other minerals may have been passing under the same name, although the only feature they had in common with topaz was the colour. The best-known example in this category is citrine, a variety of quartz.

Quartz, var. agate, cut and polished section
Rio Grande do Sul (Brazil)
Size   16 cm × 14 cm × 4.5 cm

Topaz ranges in colour from pure gold
yellow and reddish brown—the commonest
varieties—to pale blue, occasionally inter-
mingled with sea-green, and there are even
some that are pink and wine-red, the rarest
and most beautiful gemstones being of this
variety.

The colour is thought to be due to iron
oxides, and traces of chromium to produce
the pink tone.

The finest topazes are found in Brazil. The
large, gold-coloured crystals from the Ouro
Prêto district are converted by artificial heating to the favourite, rose-red gemstones.
Blue, iridescent green, and colourless stones
can be found in lapidary workshops all over
the world.

There are several outstanding deposits of
topaz in the Soviet Union. There was one
famous locality in the neighbourhood of
Mursinka in the Urals where superb, pale
blue, water-clear topazes with perfect crystal
form were found. The distinguishing feature
of the magnificent specimens in collections
is that topaz and smoky quartz occur intergrown in spherical masses in association with
albite and with reddish, iridescent lithium
mica. Other localities in the Urals region are
at Miass, where colourless crystals are found
with amazonite, and reddish pink to violet-
coloured topazes occur in the former gold
placer deposits of Sanarka. In the western
Ukraine, too, large, brown topaz crystals are
found in Wolhynia. These are associated with
very dark brown smoky quartz and beryl.

Small topaz crystals are to be found in the
Mourne Mountains in Ireland. Crystals of
large size are found at Fossum, near Modum in
Norway, and at Finbo near Falun in Sweden.
Large, water-clear topazes are also found in
large numbers in Japan and Namibia.

Precious Opal

Before the discovery of the rich opal fields in
Australia in the eighties of the last century,
the chief known locality of precious opal was
Červenica in Czechoslovakia. The origin of
the name, however, points to India, the fa-
bled land of gemstones. In Sanskrit the name
upala means 'stone' or 'gemstone,' but it probably did not refer to opal at all, for finds of this
enchanting gemstone were very rare indeed
there. Even in antiquity, opal was considered
one of the most costly of gems because of its

fascinating play of colours, mingling the red of garnet, the purple of amethyst, the golden yellow of topaz, the blue of sapphire and the green of emerald into a blaze of rainbow colours. This opalescence is not caused by inclusions, or by traces of heavy metals, but by the diffraction effects of light on the submicroscopically small planar structure of microcrystalline crystallites of cristobalite.

The typical form of the amorphous mineral is a solidified gel. It is always of secondary origin and occurs as a vein filling in sedimentary rocks. Opals, therefore, are often to be found near the surface of the Earth.

They have been surrounded with superstition from earliest times. Revered in the Orient as the stone of hope and purity, offering protection against disease, in Europe, by contrast, it came to be regarded as an unlucky stone. Queen Victoria refused to be deterred herself by the alleged magical powers of opal, and selected it as an ornamental stone for the Crown Jewels. Her decision was doubtless influenced by the finds made in Australia, which then formed part of the British Empire.

Other known localities are in Mexico (near Querétaro), Honduras and California, United States of America. Červenica in Czechoslovakia is only of historical interest, for the veins are now completely exhausted. The finest specimens from Červenica can be seen in the Treasury of the Vienna Hofburg. There is also a small collection in Košice.

Precious opal is not the only mineral in this group, even if the following varieties are less important. Fire opal was brought from Mexico by ALEXANDER VON HUMBOLDT. Fire-red to hyacinth in colour, the raw material found in Zimapan was cut and polished into the most exquisite gems. Further localities are Zabkovice and Szklary in Poland as well as deposits in New Caledonia. A new occurrence has recently been discovered in Tanzania. More common is milky opal, mostly yellowish white in colour. It becomes cloudy on dehydration but immersion in water will temporarily restore the opalescence typical of precious opal.

In wood opal the structure of fossil wood has been retained by deposits of silica replacing the wood but retaining the structure, often in great detail. The presence of heavy metals may result in many different shades of colour. Specimens from Hungary and Czechoslovakia are to be found in many collections.

Kunzite

Kunzite was discovered in 1903 near Pala in California, and was named after the American gemmologist, G F KUNZ. Its lovely pink or lilac colour and its hardness made it a highly prized gemstone. The pleochroism is a characteristic feature of kunzite and is a useful distinguishing property (see under tourmaline). In kunzite the colour is deeper along the length of crystals than across them. Its good cleavage makes cutting and grinding extremely difficult, and only the most experienced and expert craftsmen are able to work the mineral. A further characteristic of kunzite is the frequent occurrence of etched

206

faces and therefore the number of perfect display specimens is comparatively small.

The largest, twinned crystal, weighing an impressive 7.5 kilos and of quite exceptional colour, was found in 1961 in Itambacuri in Minas Gerais (Brazil). A further source of kunzite is the Antsirabe deposit on the Island of Madagascar.

Rubellite

Pink-coloured rubellite is one of the rarer varieties of tourmaline, the colour being finest in transparent crystals. The name is derived from the Latin, *rubellus*, meaning 'reddish.'

Like other varieties of tourmaline mentioned earlier, the colour and hardness of rubellite determined its selection as a gem. Yet stones of a single colour are none too common, for rubellites occur mostly with mixed colours blending into each other without any definite boundaries. These specimens with different shades of colour make eye-catching exhibits in the showcases of public collections. Red and green is a common colour combination. The largest crystals in these shades come from Pala in California. Almost equivalent specimens have been found in Governador Valadares in Brazil. There are further occurrences that have yielded rubellite of excellent quality in Ligonha in Mozambique and Antandrokomby on the Island of Madagascar.

Quartz, var. agate
Rio Grande do Sul (Brazil)
Size: 30cm × 12cm × 2cm

The Greeks and Romans described lazurite as opaque sapphire with gold spots, thus identifying its two essential features—the deep blue colour and the inclusions of pyrite. In some minerals, for instance azurite and chalcanthite, the intense blue colour is due to the presence of copper, but in lapis lazuli it results from the bonding relationships of the sulphur in the structure. Because it is the dominant property, reference to the blue colour became part of the name of the mineral. *Lapis*, meaning 'stone', comes from the Latin, while *lazaward* is the word used by the Arabs for the sky and its various shades of blue.

The oldest deposit lies north of the Hindu Kush mountains, in Afghanistan. This prized material, which occurs in association with limestone, is extracted by heating the rock and subsequently chilling it with cold water, thereby detaching large blocks from the bed rock, to be further reduced by the same means so as to make them workable by hand. There are three varieties, distinguished according to colour: neeli, indigo blue, asmani, light blue, and suvsi, green. There is a further, very rich locality which lies south of Lake Baikal in the Siberian region of the Soviet Union near the Sludyanka and Bystraya rivers.

Lapis lazuli from the Chilean Andes is intimately intergrown with limestone, and is consequently less suitable for use in ornamental jewellery.

Thousands of years ago lapis lazuli was held in high esteem by the Egyptians. They made necklaces, bangles and signet rings from it. To this day it still remains popular. Proof of this lies in the many valuable vases, bowls and objets d'art to be found in collections the world over.

Quartz, var. agate with infilled fissures in porphyry
St Egidien near Glauchau, Saxony (G.D.R.)
Size: 8 cm × 8 cm × 2.5 cm

Quartz, var. agate (polished section)
Uruguay
Size: 8 cm × 7 cm × 1.5 cm

Turquoise

In antiquity Persia had access to rich turquoise deposits, and from Persia it reached Europe by way of Turkey. The crusades were a major contribution to its wider distribution.

In ancient Egypt turquoise was mined in the Sinai Peninsula by slaves, and made into jewellery, examples of which have been discovered in the tombs of the Pharaohs. Turquoise is a secondary mineral usually found filling joints and fissures in aluminous rocks that have undergone alteration. Its soft, blue colour is produced by copper. The presence of iron produces the less popular shades of bluish to apple-green.

Until the eighteenth century turquoise was thought to be the coloured bones and teeth of mammals, and was termed odontolite. Even G AGRICOLA alluded to the fact that no other gemstone is as easy to imitate as turquoise. To this day, substantial quantities of ivory artificially tinted blue are passed off as so-called 'bone turquoise.'

The best known localities yielding excellent, gemstone quality turquoise are Nishapur in Iran, Los Cerillos in New Mexico, United States of America, a deposit already worked by the Aztecs, and the recent discoveries in the Gilewa Mountains in Tanzania.

Amber

Amber occupies a special place in the classification of minerals since, being a fossil resin, it is an organic substance. The richest and longest known locality is the so-called 'Blue Earth,' an early Tertiary (late Eocene) deposit exposed along the Baltic coast in the Kaliningrad area of the Soviet Union. The amber-bearing layers have been repeatedly disturbed, and especially by advance of the ice during the Pleistocene glaciation, so that the distribution of the amber coincides with the southern limit of the ice. Amber is often washed from exposed rocks by the sea and, because of its low density, is transported considerable distances by waves and currents.

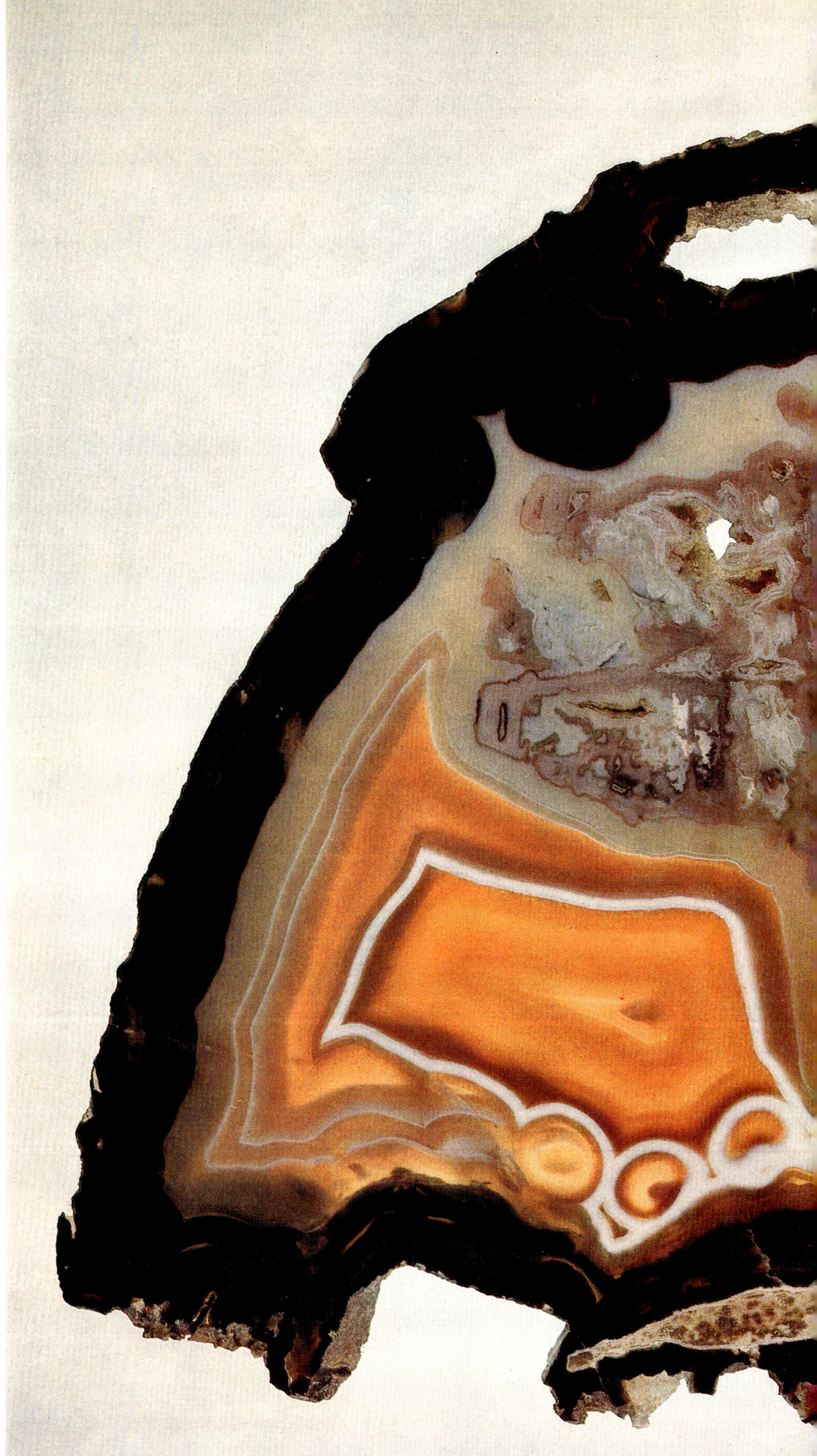

Quartz, var. agate (polished section)
Uruguay
Size: 33 cm × 21 cm × 1 cm

Such 'sea amber' is found on the coasts of the German Democratic Republic and the Federal Republic of Germany, Norway, Holland and even eastern England.

Amber from the Baltic region has been known since antiquity. It was used in medicine, in perfumes, and as jewellery. The Teutons called it *glaesum* (glass). In the thirteenth century the name Börnstein appeared in northern Germany, a reference to the burning of the stone. The Greeks used their word, *elektron*, to denote the powers of attraction of amber, because when rubbed it becomes electrically charged which can be demonstrated by picking up small pieces of paper with a piece of amber so charged.

Amber is amorphous and has no ordered internal structure. The forms of amber are extremely diverse; it occurs mostly as irregular lumps, grains, blobs or rounded discs. The largest specimen found in the Baltic region weighed almost 7 kilos. The colour varies between honey-yellow, hyacinth red and brown, blue and green being rarer. The cloudiness of amber caused by the inclusion of small air bubbles, has a considerable effect on the colour. The perfectly transparent, deep yellow pieces, with a vitreous lustre are those favoured for making into jewellery.

The weathering of amber is accelerated by exposure to air, and takes the form of a cloudy, dark crust. The thickness of the weathering crust varies depending on the locality from which the amber comes. Amber washed up on the sea shore has a very thin weathered crust because of the abrasive effect of wave action, and material from the 'Blue Earth' is fairly evenly weathered, only the core remaining unaffected. Pieces from Pleistocene strata which have been exposed to the air are extremely altered, and are a lovely brown colour.

Amber is also found in Miocene lignite deposits in Jutland.

The value and interest of amber are increased by the wonderful state of preservation of insects, lichens, mosses and ferns trapped within it.

Quartz, var. agate (banded agate)
with infilled fissures
Schlottwitz, Erzgebirge (G.D.R.)
Detail

Labradorite

Some members of the feldspar family are strikingly ornamental. The bluish-green iridescence of labradorite has made it valuable for use as an ornamental stone and in jewellery. Depending on the angle at which the light catches it, a rich iridescent play of colours appears on the surface, which is otherwise ashy-grey. The play of colours results from interference and refraction of the light on twin lamellae. Frequently the colouring is not evenly distributed over the labradorite, but forms bands of differing colours ranging from blue through green to yellowish orange. Now and again the different colours have a concentric arrangement.

Labradorite was discovered in 1770 on the coast of Labrador, by a missionary who saw its brilliant colours gleaming in the sunshine. It is this locality that gave the mineral its name. A further occurrence was discovered in 1939 by the Finnish mineralogist Latakari, near Ylijärvi. Here the labradorite glitters in all the colours of the rainbow and is consequently also known as spectrolite. Large labradorite crystals occur in the gabbro-norite of Zhitomir, in the neighbourhood of Kiev, Soviet Union, and when cut and polished, the dark rock shows good schiller effect.

Apart from labradorite, other feldspars that also possess gem qualities are orthoclase, moonstone, amazonstone and sunstone. Limpid, rich yellow orthoclase from the Island of Madagascar has recently become very popular, and is occasionally used as jewellery. After cutting, blue-white moonstone displays a wavy shimmer reminiscent of moonlight. Most moonstone comes from Sri Lanka.

Amazonstone is a variety of the potassium alkali feldspar microcline coloured green. The colour may be due to the presence of fluorine but this is not certain. ALEXANDER VON HUMBOLDT records that on his journey to South America, he was given green stones by the Indians on the Rio Negro, which were known as Amazon stones. They do not take their name from the River Amazon in South

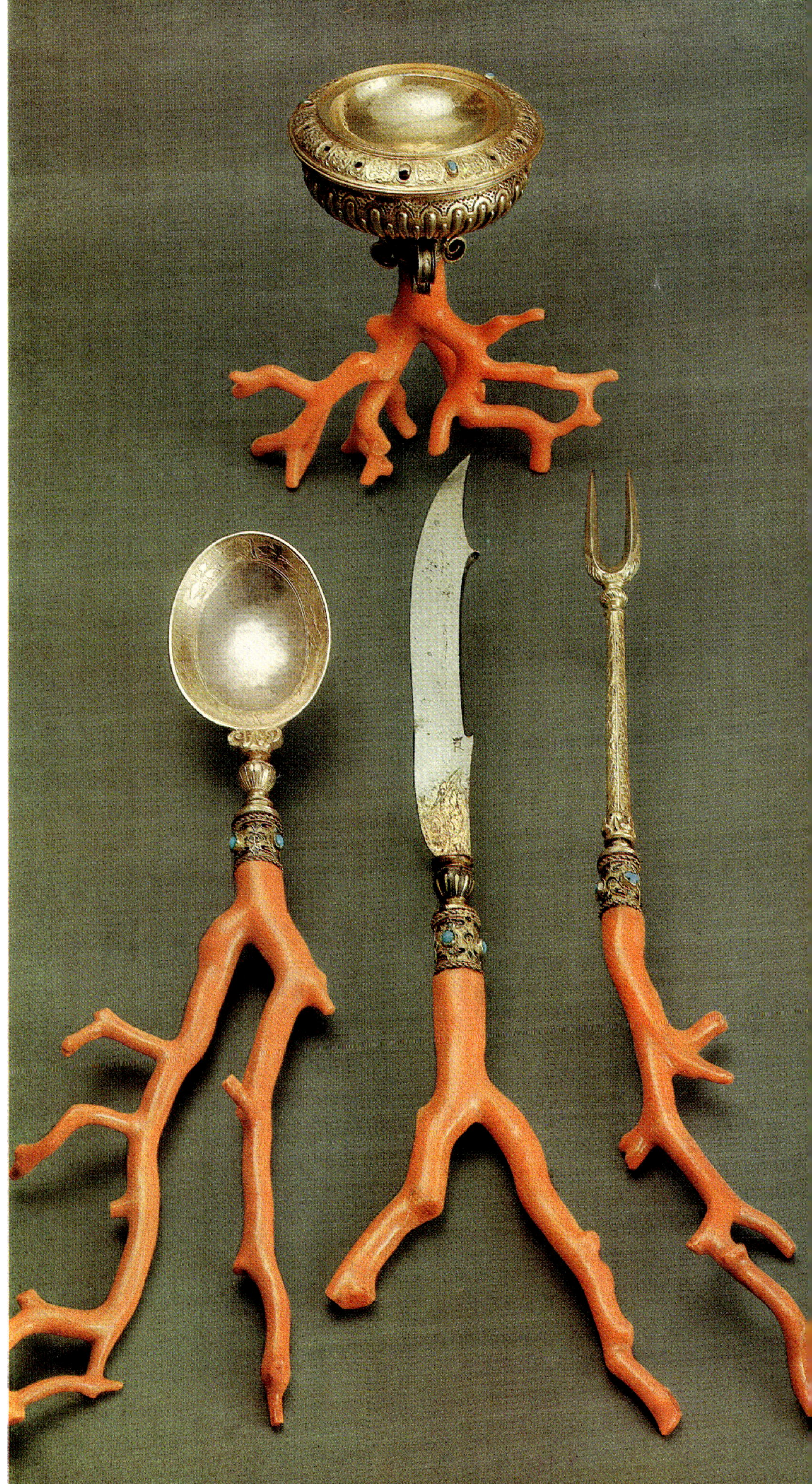

Coral cutlery (16th century)
Steel, silver gilt, turquoise
Staatliche Kunstsammlungen Dresden,
Grünes Gewölbe

213

America, but from the Indian country of Amazonas. There are further occurrences in the Ilmen Mountains in the Soviet Union, and at Pikes Peak in Colorado, United States of America.

Like labradorite, sunstone has a red schiller that depends on the angle at which the light strikes the specimen. It is produced by minute crystals and plates of hematite and goethite. The finest specimens of this particular variety of feldspar come from Tvedestrand in Norway.

Amethyst

The most popular variety of quartz is violet to lilac-coloured amethyst. The colour, which in some specimens fades on exposure to light, is ascribed to traces of iron or manganese or to long exposure to cosmic radiation. The colour of amethyst can be changed to yellow or a pale, reddish brown by heating. These 'burnt' amethysts are mostly passed off as citrines and offered as 'gold topazes' in the jewellery trade.

Amethyst has been a very popular semiprecious stone since antiquity and has always been in great demand. The name means 'unintoxicated'—the translation of the Greek word, *amethystos*—because amethyst was thought to have the power of preventing drunkenness.

The specimens most sought by collectors are those from Guanajuato and Guerrero in Mexico, comprising magnificent groups of single crystals which frequently show subtle variations in colour. The lower part of the prism is a cloudy milky colour, which passes into a colourless part, followed by a faint pink tint which intensifies into the deep violet tip of the crystal.

The most productive localities for amethyst are in Bahia and Rio Grande do Sul in the north and south of Brazil, respectively.

Order of the Golden Fleece (1744)
Three Bohemian garnets, brilliant-cut
diamonds, gold and silver
Staatliche Kunstsammlungen Dresden,
Grünes Gewölbe

Many valuable amethysts are also mined in
neighbouring Uruguay. These two countries
also supply the deep, evenly coloured material for the many gemstones that are either
facetted or cut as cabochons.

The deepest violet of all amethysts are
those that come from Zambia. Rarer items
in collections are amethysts from the Zillertal
in the Tyrol, Austria, which often have the
form of sceptre quartz, and only slightly less
beautiful than these are the zoned almost
pinky-violet specimens from the old locality
of Banská Štiavnica in Czechoslovakia.

Massive vein amethyst has been mined at
localities in the area of Puy-de-Dôme in
France, and along the Bay of Fundy, Nova
Scotia, Canada where it has been cut into
slabs for ornamental use. On the north shore
of Lake Superior in Canada, at Thunder Bay,
large crystals up to 5 or 6 inches across occur as groups associated with calcite, fluorite
and baryte in sulphide veins. Amethyst is
found at many localities in the United States
of America including fine crystals associated
with colourless and smoky quartz in the
neighbourhood of Media, Delamore County,
Pennsylvania. Gem material has been obtained at various places in North Carolina,
in Virginia and in Georgia. Amethyst is
found in the hollow trunks of petrified trees at
Amethyst Mountain in Yellowstone National
Park, Wyoming.

Order of the Golden Fleece (18th century)
Brilliant-cut diamonds, topaz, gold and silver
Staatliche Kunstsammlungen Dresden,
Grünes Gewölbe

215

Agate

Agate is a variety of chalcedony, which consists of silica together with some water. Normally chalcedony is homogeneous, but the variety known as agate is characterized by thin bands of different colours. The bands are usually arranged in concentric layers within a spherical to sub-spherical mass known as a 'geode.'

Agates, which were known to the ancient Greeks and Romans, are supposed to derive their name from the place where they were first found, the River Achates in Sicily. They differed in form from the agates we know today. PLINY described agates as brightly-coloured stones with patches, stripes or dendrites as well as patterned designs. In particular, he singled out the agates from India, as showing particularly well the features he described. Indian moss agate is still highly prized and sought after today. It is not banded, but is chalcedony which contains inclusions of what look like plants but in fact are masses of manganese oxide and it appeals greatly to the imagination, for the dark green dendrites occur in many different forms.

The largest deposits with the richest yields are in South America. Substantial quantities of rough agates are collected by the local inhabitants from the river gravels in Rio Grande do Sul in Brazil and in the northern territory of Uruguay. They reach the market in large quantities, but only a small amount is destined for jewellery and decoration, most

Three jewelled pendants
Gold, diamonds, rubies, emeralds, enamel
Left: stag with lute player (16th century)
Centre: coat of arms of the electors of
Saxony
Right: Monogram 'A' (16th century).
The letter 'A' refers either to Augustus,
the Elector of Saxony, or to his wife, Anna.
Staatliche Kunstsammlungen Dresden,
Grünes Gewölbe

Star of the Polish White Eagle
(18th century)
Brilliant-cut diamonds, rubies, gold and silver
Staatliche Kunstsammlungen Dresden,
Grünes Gewölbe

being used for a variety of industrial purposes, for example in the manufacture of
knife edges and bearings for chemical balan-
ces. Agate geodes from Mexico are also very
popular, especially if, in addition to the banded agate, they are lined with lilac-coloured
amethyst or with fine calcite crystals.

German agates are red and consequently
easily distinguished from those from other
regions, which, for the most part, have grey
and white bands. The monotonous colouring
of the South American specimens is frequently enhanced by being artificially coloured black, green or red.

Agates and other varieties of chalcedony
are obtained at many places on the Deccan
Plateau of India. Important localities are at
Ratnapura on the lower Narbuda River and
in the area north of Rajkot on the Kathia-
war Peninsula. Agate has also been produced
commercially in the Malagasey Republic and
is found in many regions of the Soviet
Union. Among the numerous localities in
the United States of America, may be mentioned the Keweenaw Peninsula, Michigan,
around Paterson, New Jersey, and a wide
area of South Dakota.

Petrified Wood

Petrified wood is half mineral, half fossil, the
silicification contributing to the preservation
of the wood. The structure of the organic
matter, the cell walls, aerial roots and annual
rings can frequently still be recognized in
detail, as a result of impregnation by silica,
even though some of the organic matter was
dissolved. Seen thus, the process is a typical
example of fossilization.

Any account of the mineralization of petrified wood must include mention of chalcedony, agate, jasper and opal. The core of the
wood, and frequently also major cracks and
fissures are here and there replaced by or infilled with beautifully banded, variegated
agates. Even the presence of fluorite has been
confirmed in preserved tree trunks.

Hat jewel (18th century)
Brilliant-cut diamonds and silver
Staatliche Kunstsammlungen Dresden,
Grünes Gewölbe

217

A G WERNER was the first mineralogist to
include fossil wood in the systematic classification of minerals.

Petrified wood occurs locally in large
quantities; sometimes it is appropriate to
speak in terms of 'fossil forests.' The most important examples of these flora are Calamites
and conifers such as Araucaria.

In the Virgin Valley, Nevada, petrified
wood replaced by both chalcedony and opal
occurs. It is also found in the Petrified Forest
National Park, Arizona and the Yellowtree
National Park.

Jasper

Jasper, like agate, is a variety of chalcedony.
However, unlike agate, jasper varieties are
completely opaque.

They are widely distributed.

Jasper has been obtained commercially
from Sicily, the German Democratic Republic, India, the Soviet Union, Mexico and the
United States of America. Pebbles of brown
and red jasper are common in the deserts of
North Africa and elsewhere. Green and banded red and green jasper suitable for ornamental purposes, are found at many localities in
the Ural Mountains and the Altai Mountains
of the Soviet Union and in the Erzgebirge in
the German Democratic Republic. There are
numerous localities of jasper showing excellent colours in Brazil, India, the Soviet
Union, Italy, Mexico, and in the United
States of America including areas of Texas,
Arizona, California, and Rhode Island.

Scent bottle (16th to 17th century)
Onyx agate
Staatliche Kunstsammlungen Dresden,
Grünes Gewölbe

Cabinet; Life's greatest joy
Detail (18th century)
Gold, silver, gems, agate, horn, pearls,
enamel
Staatliche Kunstsammlungen Dresden,
Grünes Gewölbe

Florentine mosaic (18th century)
Amethyst, agate, jasper, chalcedony,
lapis lazuli
Staatliche Kunstsammlungen Dresden,
Grünes Gewölbe

Circular box with flower still life
77 Saxon precious stones
Staatliche Kunstsammlungen Dresden,
Grünes Gewölbe

Page 222
Three spoons (16th to 17th century)
Gold, enamel, jasper
Staatliche Kunstsammlungen Dresden,
Grünes Gewölbe

Bibliography

BANCROFT, P.: *The World's Finest Minerals and Crystals.* Viking Press, Inc., New York 1973

BANK, H.: *Aus der Welt der Edelsteine.* Pinguin-Verlag, Innsbruck 1971

BAUER, M.: *Edelsteinkunde.* Bernhard Tauchnitz, Leipzig 1932

BRAUNS, R.: *Das Mineralreich.* J. F. Schreiber, Esslingen and Munich 1912

COPELAND, L. L.: *Diamonds, famous, notable and unique.* G. I. A. 1966

DE MICHELE, V.: *Il mondo dei Cristalli.* Istituto Geografico de Agostini, Novaro 1967

DESAUTELS, P. E.: *The Mineral Kingdom.* Madison Square Press, Grosset and Dunlap, New York 1968

FISCHER, W.: *Mineralogie in Sachsen von Agricola bis Werner.* Verlagsbuchhandlung C. Heinrich, Dresden 1939

GÜBELIN, E.: *Edelsteine.* Silva-Verlag, Zurich 1969

GUILLEMIN, C.: *En Visitant les Grandes Collections Minéralogiques Mondiales.* B. R. G. M., Paris 1964

HEY, M. H.: *Chemical Index of Minerals.* British Museum (Natural History)

HINTZE, C.: *Handbuch der Mineralogie,* edited by K. F. Chudoba. Walter de Gruyter u. Co., Berlin 1939

HURLBUT, C. S. JR.: *Schätze aus dem Schoss der Erde.* Bertelsmann Lexikon-Verlag, Gütersloh—Berlin 1972

KIPFER, A.: *Mineralindex.* Ott Verlag, Thun and Munich 1974

KOUŘIMSKÝ, J.: *Bunte Welt der Mineralien.* Artia, Prague 1977

KRUMBIEGEL, G. and H. WALTHER: *Fossilien.* VEB Deutscher Verlag für Grundstoffindustrie, Leipzig 1977

LIEBER, W.: *Der Mineraliensammler.* Ott Verlag, Thun and Munich 1971

LÜSCHEN, H.: *Die Namen der Steine.* Ott Verlag, Thun and Munich 1968

MENZHAUSEN, J.: *The Green Vault.* 2nd edition. Edition Leipzig 1970

METZ, R. and A. E. FANCK: *Antlitz edler Steine.* Belser Verlag, Stuttgart 1964

MUNDLOS, R.: *Wunderwelt im Stein.* Bertelsmann Lexikon-Verlag, Gütersloh—Berlin 1976

RINNE, F. and M. BEREK: *Anleitung zur allgemeinen und Polarisationsmikroskopie der Festkörper im Durchlicht.* 3rd edition. Editors: H. Schumann, F. Kornder. E. Schweizerbart'sche Verlagsbuchhandlung, Stuttgart 1973

RYKART, R.: *Bergkristall.* Ott Verlag, Thun and Munich 1971

SCHUMANN, H.: *Einführung in die Gesteinswelt.* 2nd edition. Verlag von Theodor Steinkopff, Dresden and Leipzig 1957

WOOLLEY, A. R., A. C. BISHOP and W. R. HAMILTON: *The Hamlyn Guide to Minerals, Rocks and Fossils.* Hamlyn, London 1974

Freiberger Forschungsheft C 223: Abraham Gottlob Werner. VEB Deutscher Verlag für Grundstoffindustrie, Leipzig 1967

World Directory of Mineral Collections. Commission on Museums of the International Mineralogical Association, 1974

Leningradsky Gorny Institut 1773–1973. Vysshaya Shkola, Moscow 1973

Fundgrube, Zeitschrift für Geologie, Mineralogie, Paläontologie, Speläologie. Editor: Kulturbund der DDR, ZFA Geowissenschaften, Berlin

Der Aufschluss. Zeitschrift für die Freunde der Mineralogie und Geologie. Editor: V. F. M. G. e. V., Heidelberg

Lapis. Monatsschrift für Liebhaber und Sammler von Mineralien und Edelsteinen. Christian Weise Verlag, Munich

224

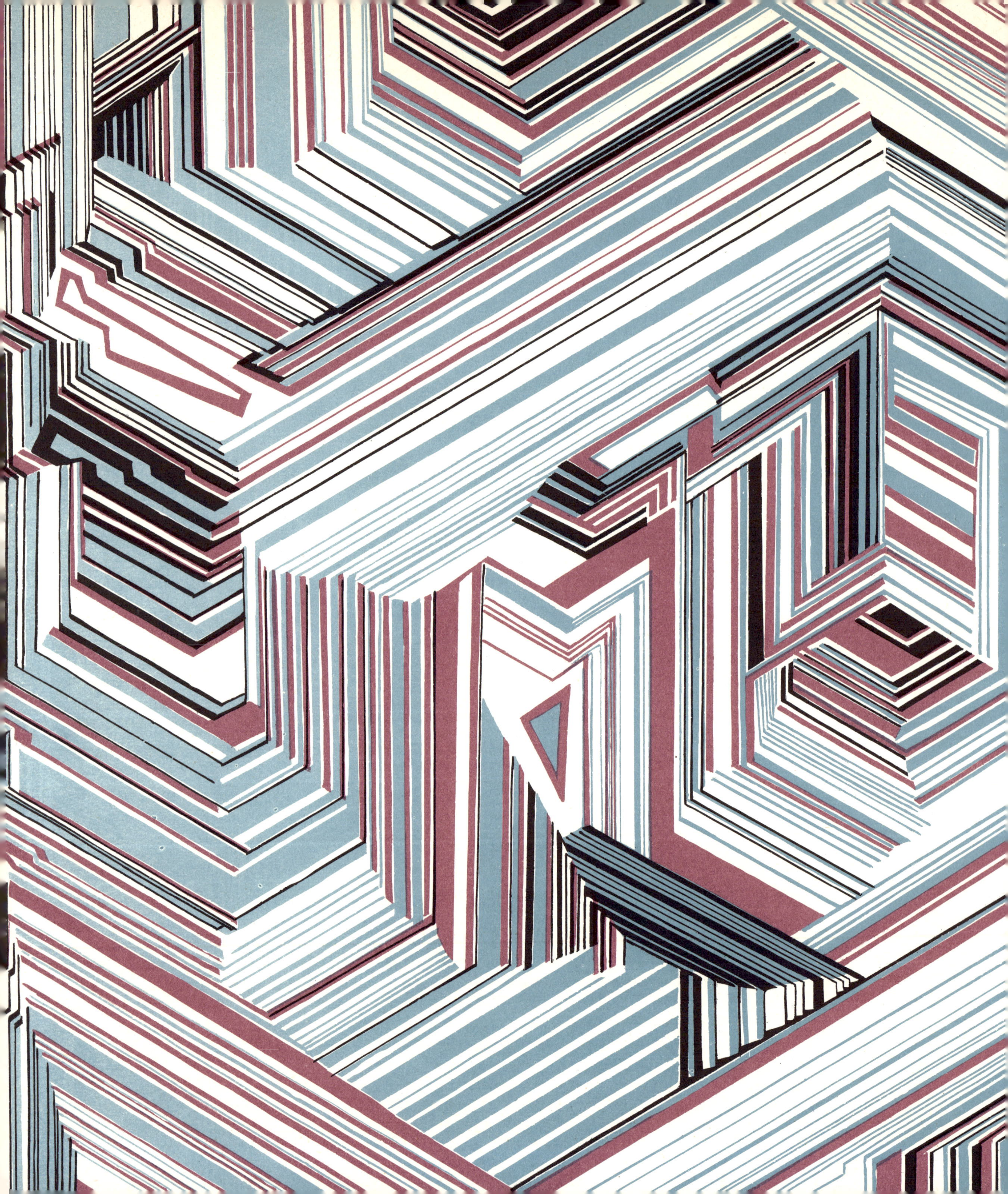